Microelectronics and Microsystems

Springer
London
Berlin
Heidelberg
New York
Barcelona
Hong Kong
Milan
Paris
Singapore
Tokyo

Luigi Fortuna, Giuseppe Ferla
and Antonio Imbruglia (Eds)

Microelectronics and Microsystems

Emergent Design Techniques

With 212 Figures

Springer

Luigi Fortuna, Professor
Dipartimento Elettrico Elettronico e Sistemistico, Università degli Studi di Catania, Viale Andrea Doria 6, 95125 Catania, Italy

Giuseppe Ferla, Professor
Antonio Imbruglia, Doctor
STMicroelectronics, Stradale Primosole 50, 95121 Catania, Italy

ISBN-13: 978-1-4471-1174-0 e-ISBN-13: 978-1-4471-0671-5

DOI: 10.1007/978-1-4471-0671-5

British Library Cataloguing in Publication Data
Microelectronic and microsystems : emergent design
techniques
1.Microelectronics 2.System design
I.Fortuna, L. (Luigi), 1953- II.Ferla, Giuseppe
III.Imbruglia, Antonio
6213'81

Library of Congress Cataloging-in-Publication Data
A catalog record for this book is available from the Library of Congress

Springer-Verlag London Berlin Heidelberg
a member of BertelsmannSpringer Science+Business Media GmbH
http://www.springer.co.uk

Softcover reprint of the hardcover 1st edition 2002

Typesetting: Electronic text files prepared by editors

69/3830-543210 Printed on acid-free paper SPIN 10836550

To all people that
day by day
encourage our work

"To be a man means constant revision like correcting a writing"
Chinese Proverb

Preface

The book presents the best contributions, extracted from the theses written by the students who have attended the second edition of the Master in Microelectronics and Systems that has been organized by the Universitá degli Studi di Catania and that has been held at the STMicroelectronics Company (Catania Site) from May 2000 to January 2001.

In particular, the mentioned Master has been organized among the various activities of the "Istituto Superiore di Catania per la Formazione di Eccellenza". The Institute is one of the Italian network of universities selected by MURST (Ministry University Research Scientific Technology).

The first aim of the Master in Microelectronics and Systems is to increase the skills of the students with the Laurea Degree in Physics or Electrical Engineering in the more advanced areas as VLSI system design, high-speed low-voltage low-power circuits and RF systems.

The second aim has been to involve in the educational program companies like STMicroelectronics, ACCENT and ITEL, interested in emergent microelectronics topics, to cooperate with the University in developing high-level research projects.

Besides the tutorial activity during the teaching hours, provided by national and international researchers, a significant part of the School has been dedicated to the presentation of specific CAD tools and experiments in order to prepare the students to solve specific problems during the stage period and in the thesis work.

We would like to remark that the stage themes and the thesis subjects have been accurately chosen in accordance with some specific emerging topics and we would like to thank both the tutors of the University, and those of companies who have encouraged and stimulated the activity of the students. The students, in fact, have reached a high level of preparation and have often presented, in their theses, innovative solutions. The main results are reported in this book.

The material collected in this book consists in the more recent results in microelectronics regarding a wide spectrum of topics from technology to system, from the device to the circuit design.

The book will be diffused both at the University and at the RD divisions of the Etna Valley (CT) companies that are evolved in emergent projects in microelectronics.

To conclude we would like to warmly thank Prof. Giovanni De Micheli both for his attracting lectured during the course and for having accepted in his laboratories at the Stanford University some of the Master students to develop their final work. Our thanks are also both for the STMicroelectronics Staff and for the Department of Electrical Electronics and System Engineering staff of the Universitá degli Studi di Catania that allowed us to conclude successfully the initiative.We appreciated the work of Dr. M.A. Gemma during the preparation of this volume.

Prof. G. Ferla
DSG R & D Director,
STMicroelectronics S.r.l.
Italy

Prof. L. Fortuna
IEEE Fellow
Istituto Superiore di Catania per la Formazione di Eccellenza.
Universitá degli Studi di Catania
Italy

Dr. A. Imbruglia
DSG R & D Rad-Hard Product Development Manager,
STMicroelectronics S.r.l.
Italy

Contents

BiPMOS Fractional-N Frequency Synthesizer For Wideband Wireless Systems

Tino Copani

DEES, Engineering Faculty, University of Catania
Viale A. Doria 6, 95125 Catania - Italy

Introduction

This work is aimed at designing a Phase Locked Loop (PLL) circuit to be used as a frequency synthesizer in a RF transceiver. In modern wireless communication systems the local oscillator phase noise specification sets the allowed strength of nearby channels. In the receiver (Figure 1), the PLL frequency f_{lo} is set to a fixed offset from the desired channel f_{RF} to downconvert it to the intermediate frequency f_{IF}.

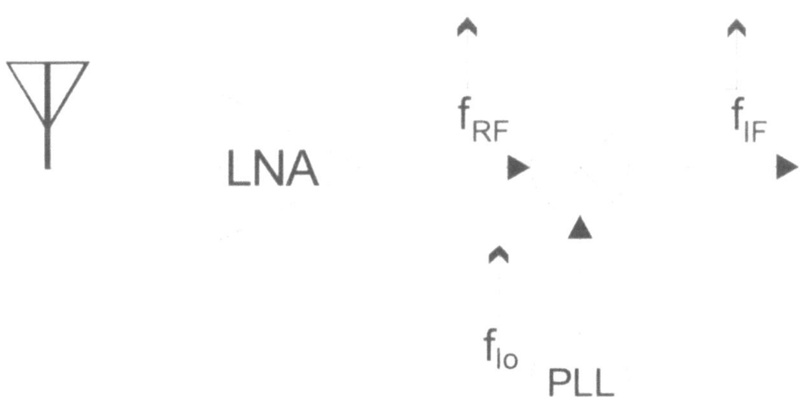

Figure 1. Superetherodyne receiver

Whenever the noise spectrum around f_{lo} is high enough, the nearby channels can be down converted to the same intermediate frequency (Figure 2).

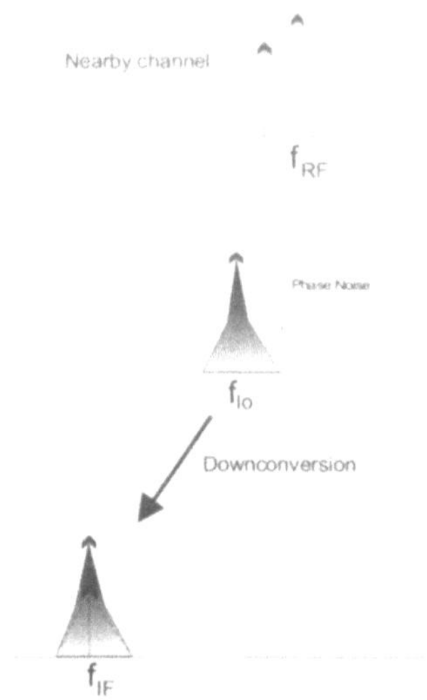

Figure 2. Downconversion of undesired nearby channel

Hence, one of the most difficult challenges in designing a PLL synthesizer is the reduction of phase noise and spur tone content.

In order to meet requirements of a wideband RF system, such as WCDMA or UMTS, the fractional-N technique has been used, that allows both better noise and spur free performance over classical designs.

Fractional-N Phase Locked Loop

In Figure 3 the simplified block diagram of a PLL is shown.

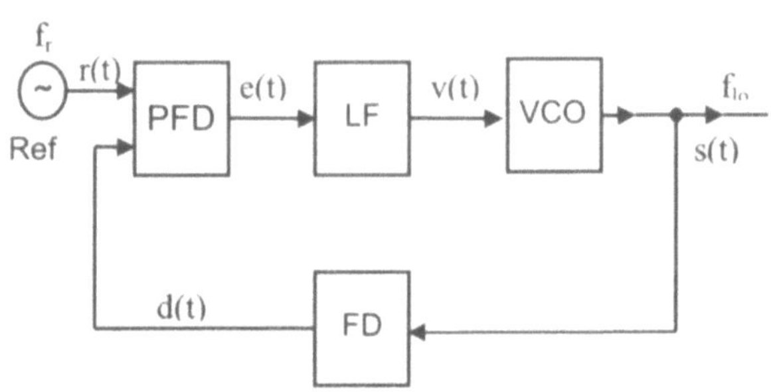

Figure 3. Phase locked loop

The desired output frequency f_{lo} is generated locking a voltage-controlled oscillator (VCO) to a crystal-derived reference (REF) by means of a phase-frequency detector (PFD). Theoretically, loop filter (LF) avoids any undesired spur signal, whereas the frequency divider (FD) makes the output frequency to be multiple of the reference frequency:

$$f_{lo} = N \cdot f_r \tag{1}$$

When N is a whole number, the reference f_r sets a limit to the PLL resolution. When a high resolution is required, e.g., very narrow channel steps are needed, then f_r has to be low and the loop band has to be narrow. This situation results in a long settling time while the VCO phase noise can't be well rejected.

To overcome this limit, a fractional-N technique (Figure 4) has been investigated. Here, at any time, a ΣΔ modulator sets the division ratio of the multi-modulus divider (MMD).

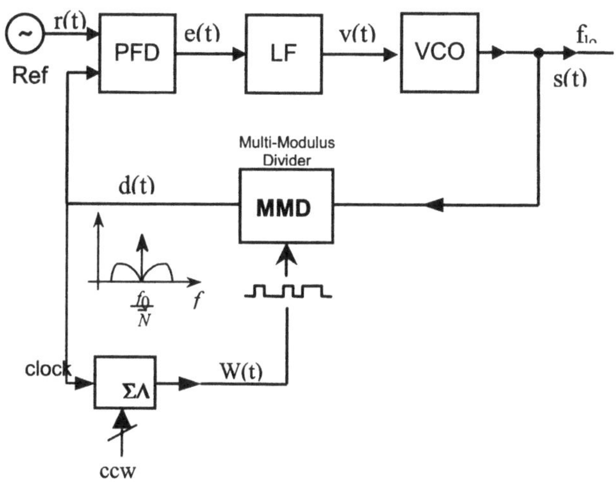

Figure 4. Fractional-N PLL

The channel control word (CCW) at the modulator input fixes the mean value $\overline{N}$ of the division ratio and hence the output frequency:

$$f_{lo} = E\{N(t) \cdot f_r\} = \overline{N} \cdot f_r \tag{2}$$

In this way, $\overline{N}$ may be a fractional number and flo can be tuned by steps smaller than fr.

With respect to Figure 4, the dynamic behaviour of the PLL follows the equation below:

$$\frac{de}{dt} + 2 \cdot \pi \cdot \frac{K_\phi \cdot K_0}{N} \cdot e(t) * h(t) = K_\phi \cdot \left(\omega_r - \frac{2 \cdot \pi \cdot f_f}{N} \right) \tag{3}$$

where $K\phi$, Ko and f_f are defined as the phase detector gain, VCO gain and free running oscillation frequency. Respectively, the $h(t)$ term is the impulse response of the loop filter (LF). When the PLL is locked then the phase difference $e(t)$ between $r(t)$

and $d(t)$ is small and the output frequency f_{lo} becomes a linear function of the reference f_r. Thus, the PLL can be modeled as the linear system shown in Figure 5.

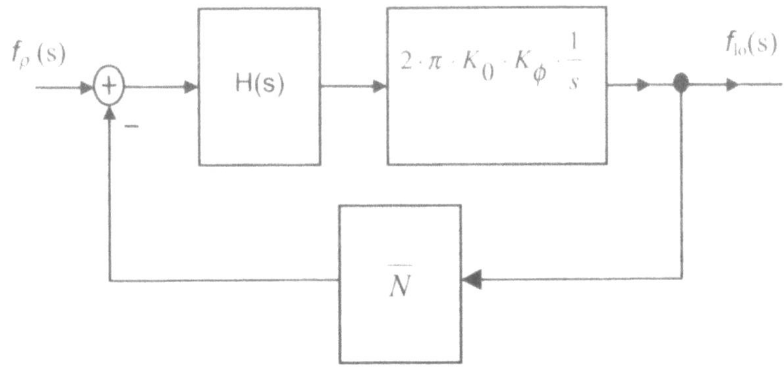

Figure 5. PLL linear model

In Figure 5, the VCO is modeled as an ideal integrator ($1/s$) while $H(s)$ is the loop filter transfer function.

In such a system, it is well known that the overall phase noise depends on the noise contribution of each block and on the loop transfer function. The noise contributors due to frequency divider and phase detector are attenuated by the loop gain outside the band of the system, whereas VCO phase noise is reduced inside the loop band as it is shown in Figure 6:

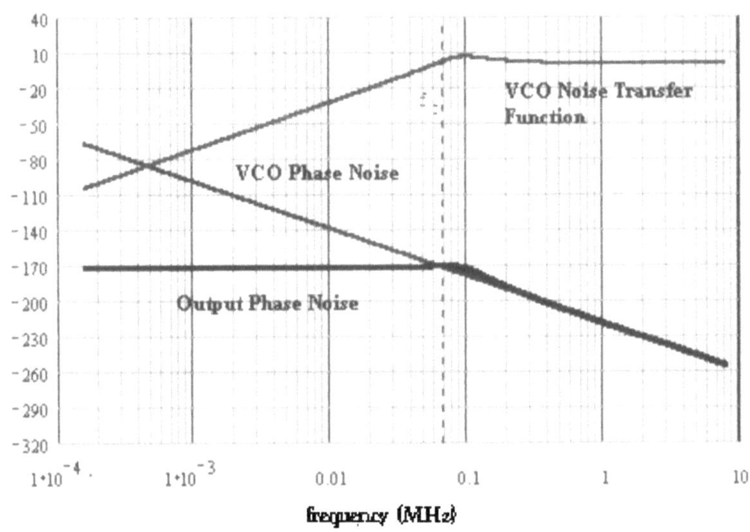

Figure 6. PLL output phase noise due to VCO

The PLL linear model of Figure 5 is useful to investigate the frequency divider contribution to the overall phase noise at the system output. The frequency divider is modeled as an ideal divider added to a zero mean value noise source:

$$N(t) = \overline{N} + n(t) \tag{4}$$

The noise $n(t)$ comes from the commutation of the division ratio in the frequency divider. When n(t) is a periodic waveform, then spur tones appear around the desired output frequency flo.

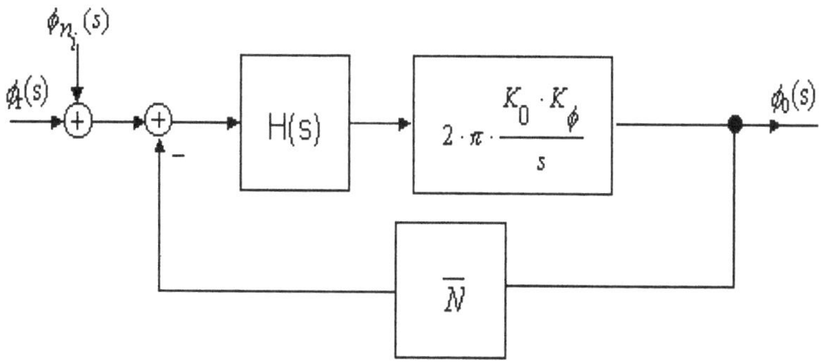

Figure 7. Frequency divider noise at the PLL input

If we take into consideration an equivalent additive noise source due to $n(t)$ at the PLL input (Figure 7), then the contribution of the frequency divider at the overall phase noise can be easily derived:

$$\phi_0(s) = \frac{2 \cdot \pi \cdot K_0 \cdot K_\phi \cdot \frac{H(s)}{s}}{1 + 2 \cdot \pi \cdot \frac{K_0 \cdot K_\phi}{\overline{N}} \cdot \frac{H(s)}{s}} \cdot \phi_{n_i}(s) \tag{5}$$

where:

$$\phi_{n_i}(s) = \frac{2\pi f_r}{s \cdot \overline{N}} \cdot n(s) \tag{6}$$

Hence, the noise power spectral density (PSD) $S_o(\omega)$ at the PLL output is given by:

$$S_0(\omega) = \left| \frac{2 \cdot \pi \cdot \frac{K_0 \cdot K_\phi}{\overline{N}} \cdot \frac{H(j\omega)}{j \cdot \omega}}{1 + 2 \cdot \pi \cdot \frac{K_0 \cdot K_\phi}{\overline{N}} \cdot \frac{H(j\omega)}{j \cdot \omega}} \cdot \frac{\omega_r}{j \cdot \omega} \right|^2 \cdot S_n(\omega) \tag{7}$$

Here, $S_o(\omega)$ is defined as the noise power spectral density of the control signal W(t) (Figure 4). Later on, it will be shown that when W(t) comes from a ΣΔ modulator, then the quantization noise $S_o(\omega)$ is shaped along the PLL band by a high pass transfer function. In Figure 8, the frequency divider noise contribution for different orders of ΣΔ modulation is plotted.

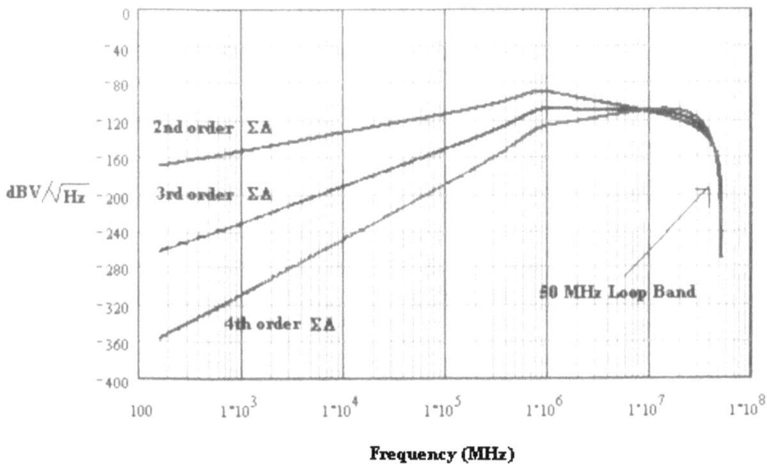

Figure 8. ΣΔ modulator contribution to in-band output noise

ΣΔ Modulators

In this chapter noise shaping fundamentals and different ΣΔ modulator topologies are presented, the solutions that better fulfill phase noise requirements are chosen for fractional-N PLL design.

As shown in Figure 9, a 1st-order ΣΔ modulator consists of a forward Euler integrator block and a quantizer block:

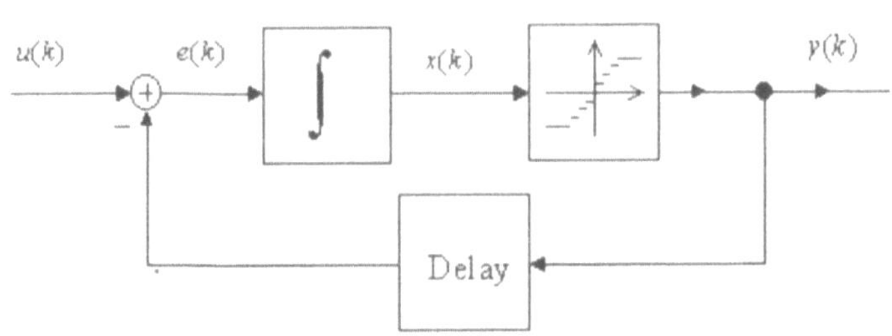

Figure 9. 1st-order ΣΔ modulator

Better to investigate the circuit behaviour, the Z-domain model of the modulator is reported in Figure 10:

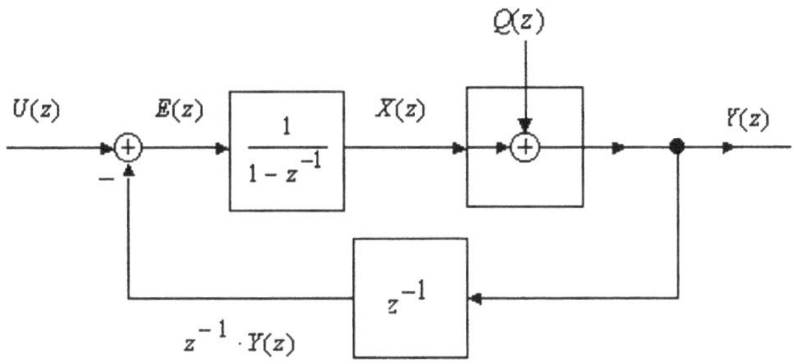

Figure 10. Z-domain 1st-order ΣΔ modulator model

Here, the quantizer block is modeled as a random noise source added at the integrator output. Clearly, we have:

$$Y(z) = U(z) + \left(1 - z^{-1}\right) \cdot Q(z) \tag{8}$$

The second term in (8) shows that the noise transfer function *NTF(f)* has a high pass behaviour in the frequency domain:

$$|NTF(f)| = 2 \cdot \sin\left[2 \cdot \pi \cdot \left(\frac{f}{2 \cdot f_s}\right)\right] \tag{9}$$

here, f_s is defined as the system sampling frequency. If we model the quantizer as a zero mean value noise source with uniform distribution in [$-f_s$, f_s], then its power spectral density $S_q(f)$ is given by:

$$S_q(f) = \frac{1}{6 \cdot f_s} \cdot \frac{V^2}{2^{2 \cdot n + 2}} \tag{10}$$

where V is the quantizer dynamic range and n is defined as the bit resolution. Combining Equations 9 and 10, the quantizer noise at the ΣΔ modulator output can be calculated. To prove the validity of the above statements, the system has been simulated with SPW CAD tool from Cadence. Simulations show an excellent agreement with the theoretical statements as long as the input is a random signal. In Figure 11, the theoretical behaviour is compared to the simulated ΣΔ modulator output when a costant signal is applied at the input.

Finally, we have to remark that the 1st-order ΣΔ modulator gives a stable output as long as the input signal doesn't overflow the quantizer dynamic range V.

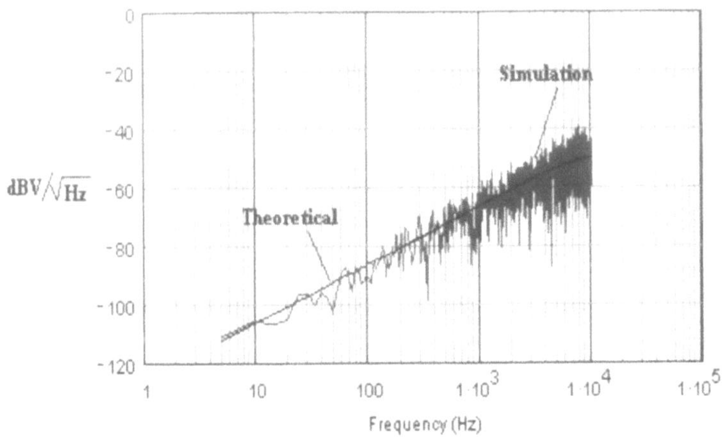

Figure 11. Quantization noise PSD at the 1st-order ΣΔ output

In Figure 12 a digital implementation of the 1st-order ΣΔ modulator is presented. As shown, the circuit consists of an n-bit full adder and an *n*-bit *D*-latch register. The word length *n* is chosen as a trade off between the required step resolution and the dynamic range of the input signal.

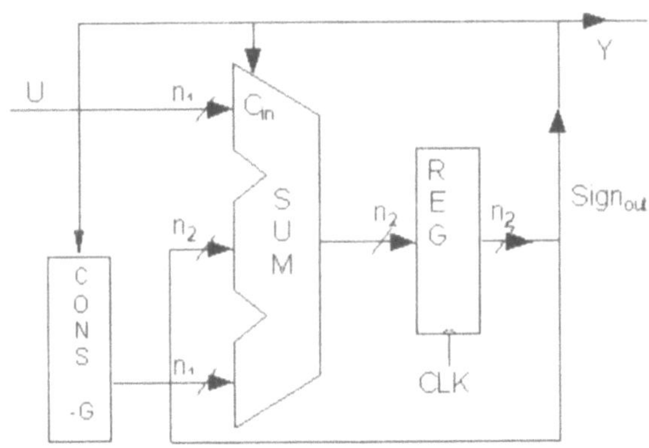

Figure 12. Digital implementation of 1st-order ΣΔ

Although the 1st-order ΣΔ modulator has a simple design, it can't be successfully employied in a fractional-N frequency synthesizer. In fact, when the input signal is a constant the output becomes a periodic waveform that plainly produces spur tones at the PLL output. Whenever a narrow loop band is used in the PLL, the spurs are attenuated yet, the system doesn't achieve the requirements of recent wideband mobile systems due to a high settling time and VCO phase noise contribution.

To overcome this problem, high order ΣΔ modulators have been investigated. In fact, it can be shown that as the order of the noise transfer function *NTF(f)* grows, the circuit outputs a more randomized signal even if a continuous signal is applied at the input. This allows, in turn, a reduction of the quantization noise and spur tones.

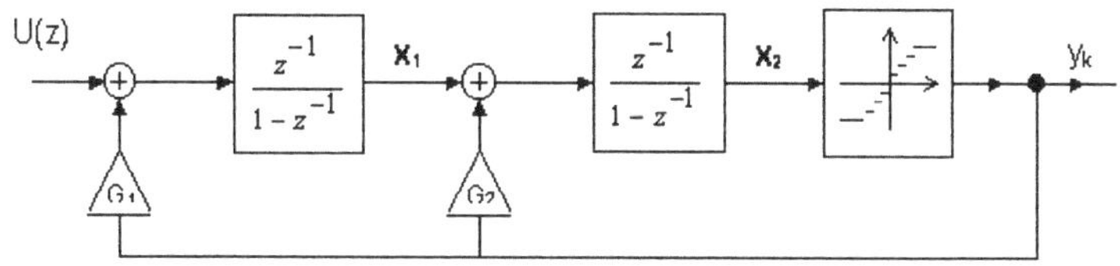

Figure 13. 2nd-order ΣΔ modulator

In Figure 13 a 2nd-order ΣΔ modulator is showed. When the fixed gain blocks G_1, G_2 are set so that $G_2=2G_1=1$, we have:

$$Y(z)=z^{-2}\cdot U(z)+\left(1-z^{-1}\right)^{2}\cdot Q(z) \tag{11}$$

Statement 11 shows that the noise transfer function has a 2nd-order high pass behaviour, allowing a better drop of in-band noise of the frequency divider than the 1st-order modulator. In general, for a Lth order ΣΔ modulator we have:

$$Y(z)=z^{-L}\cdot U(z)+\left(1-z^{-1}\right)^{L}\cdot Q(z) \tag{12}$$

hence, combining (10) and (12) we obtain the power spectral density of the quantization noise $S_n(f)$ at the modulator output:

$$S_n(f)\cong\frac{V^2}{f_s}\cdot\frac{\pi^{2\cdot L}\cdot 2^{2\cdot(L-n)-3}}{3}\cdot\left(\frac{f}{f_s}\right)^{2\cdot L} \tag{13}$$

The digital implementation of a 2nd-order ΣΔ modulator and the simulated output noise PSD are shown in Figures 14 and 15.

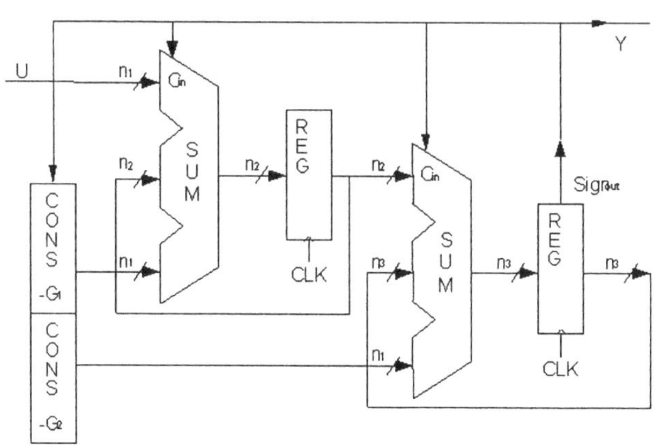

Figure 14. Digital implementation of a 2nd-order ΣΔ modulator

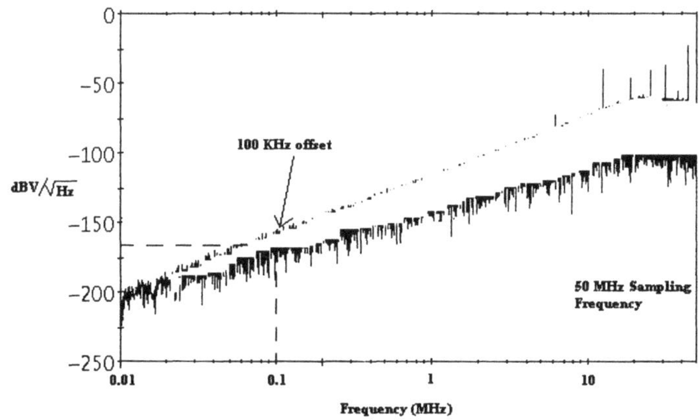

Figure 15. Quantizaton noise PSD for a 2nd-order ΣΔ modulator

In order to achieve a further drop of the in-band noise to fulfill the requirements of a new wideband wireless communication system (phase noise below *–110 dBc/Hz @ 100 KHz* offset from the carrier f_{lo}), the so called MASH ΣΔ modulators have been taken into consideration Furthermore, unlike the earlier shown topologies, a MASH modulator of the same order gives an unconditionally stable output. In Figure 16 a MASH *2-1* modulator has been reported:

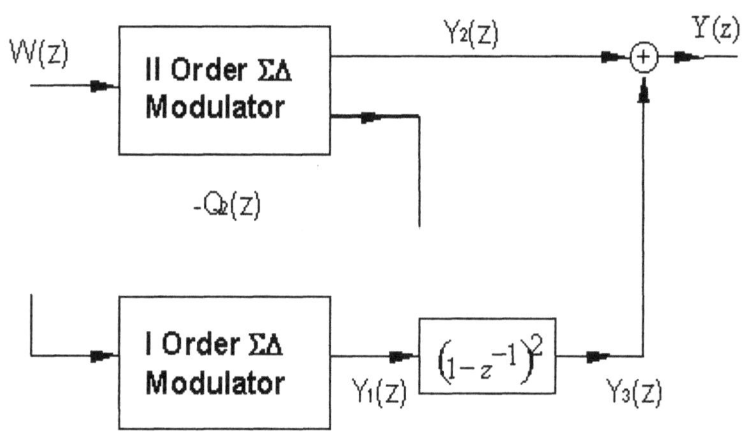

Figure 16. MASH *2-1* ΣΔ modulator

Here, the quantization noise $Q_2(z)$ at the output of a 2nd-order ΣΔ modulator feeds a 1st-order modulator and then it is perfectly removed at the system output ($Y(z)$) by the sum block. So we have:

$$Y_1(z) = W(z) + Q_1(z) \cdot \left(1 - z^{-1}\right)^3 \tag{14}$$

The MASH *2-1* modulator has been simulated and compared to a 2nd-order ΣΔ modulator (Figure 17). We can notice a decrease of the in-band noise as well as fewer high frequency spur tones.

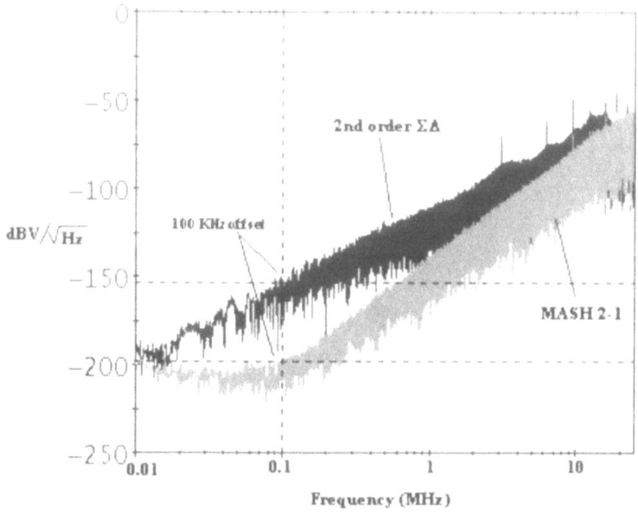

Figure 17. Noise PSD for a 2nd-order ΣΔ and a MASH *2-1* modulator

In order to completely remove those high frequency spurs, a MASH *1-1-1* modulator has been considered (Figure 18). The modulator is derived from the previous by substituting the 2nd-order ΣΔ block with two 1st-order modulators. Reported simulation of Figure 19 shows the lowest in-band noise contribution as well as a spur free behaviour.

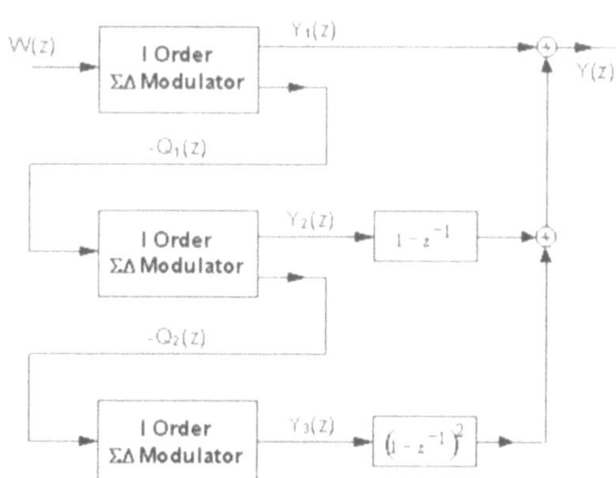

Figure 18. MASH *1-1-1* ΣΔ modulator

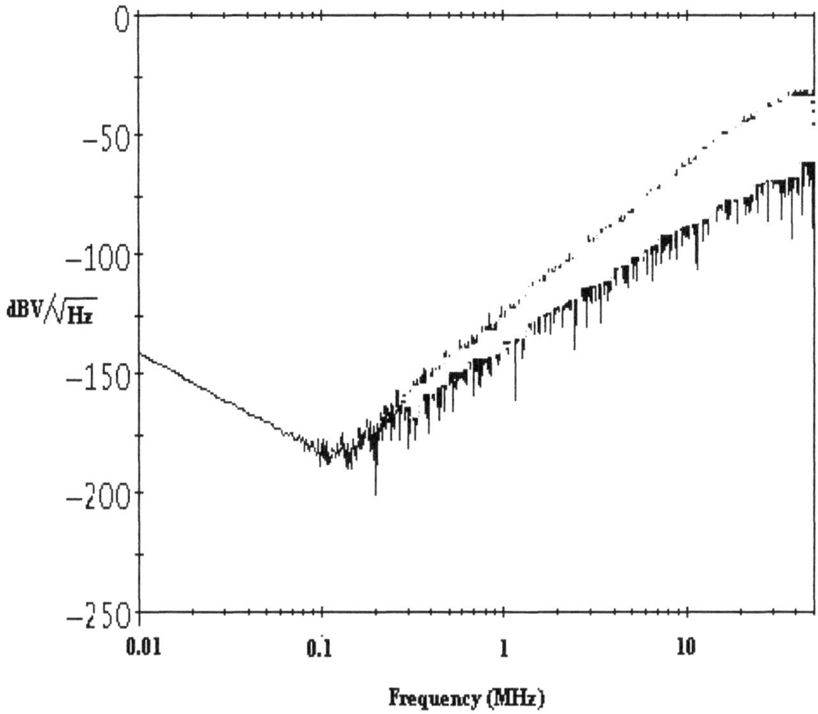

Figure 19. Noise PSD at the MASH *1-1-1* modulator output

This overall performance makes the MASH *1-1-1* as the candidate division ratio controller in a fractional-N PLL tuner.

However, MASH solutions show a main significant drawback: unlike the classical L^{th} order ΣΔ modulators, they have a parallel architecture with an n-bit long digital word as output. Therefore, while classical ΣΔ solutions (with 1-bit output) require a simple dual modulus divider (N, N+1) in the PLL system, MASH modulators require a more complex multi-modulus divider (N, N+1, …N+M).

Phase Locked Loop Design

In this chapter, general guidelines followed in PLL circuit design will be described. For the most critical blocks, circuit details will be discussed.

As stated earlier, this design should accomplish the requirements of recent wireless communications systems such as the new Universal Mobile Telecommunications System (UMTS).

The system specifications set a 2 GHz central frequency, 5 MHz channel spacing and 200 kHz channel raster. Furthermore, because of the maximum allowed adjacent channel strength, we have that the local oscillator phase noise should not exceed –110dBc/Hz at 100 KHz offset from the carrier frequency. Our goal is to demonstrate that this constraint could be fulfilled using a ΣΔ Fractional-N PLL tuner.

As seen earlier, different ΣΔ topologies may be employed, trading design complexity with noise and spur free performance.

Hence, It was preferred to implement the MASH *1-1-1* division ratio controller on a Field Programmable Gate Array (FPGA) whereas the remaining of the PLL circuit have been integrated on a recent high frequency bipolar technology. FPGA solution should guarantee design flexibility and reusability.

As MASH modulators give a n-bit long output, a multi-modulus divider was designed using differential Emitter Coupled Logic (ECL). The circuit allows any division ratio in the range 32 - 127 and it provides interfaces to and from CMOS FPGA block. From statement (1), it follows that when flo=2 GHz the allowed reference frequency is comprised in the range 15.75 MHz – 62.5 MHz.

For PLL integration was taken into consideration the BiPMOS technology provided by the STMicrolectronics plant of Catania.

This process offers a high frequency bipolar transistor (ft > 40 GHz) together with a p channel MOS transistor. The last device fills the lack of good pnp transistors typical of modern high frequency bipolar technologies, allowing a good design of interface circuits to CMOS and single ended charge pump circuits.

It's well known that the most critical block in PLLs for RF transceiver applications is the voltage controlled oscillator. In general, VCO phase noise limits PLL performance and in most commercial design, it makes use of external passive devices.

In this design a fully integrated VCO is taken into consideration, furthermore, to lower circuit sensitivity to process variations a novel structure is investigated.

All the bias voltages and currents throughout the PLL blocks are generated by band gap reference circuits for thermal stability.

At the PLL output an open collector buffer is capable of driving a 50 Ohm load (test equipment input impedance) with a –6 dBm signal.

An overview of the PLL schematic is shown in Figure 20.

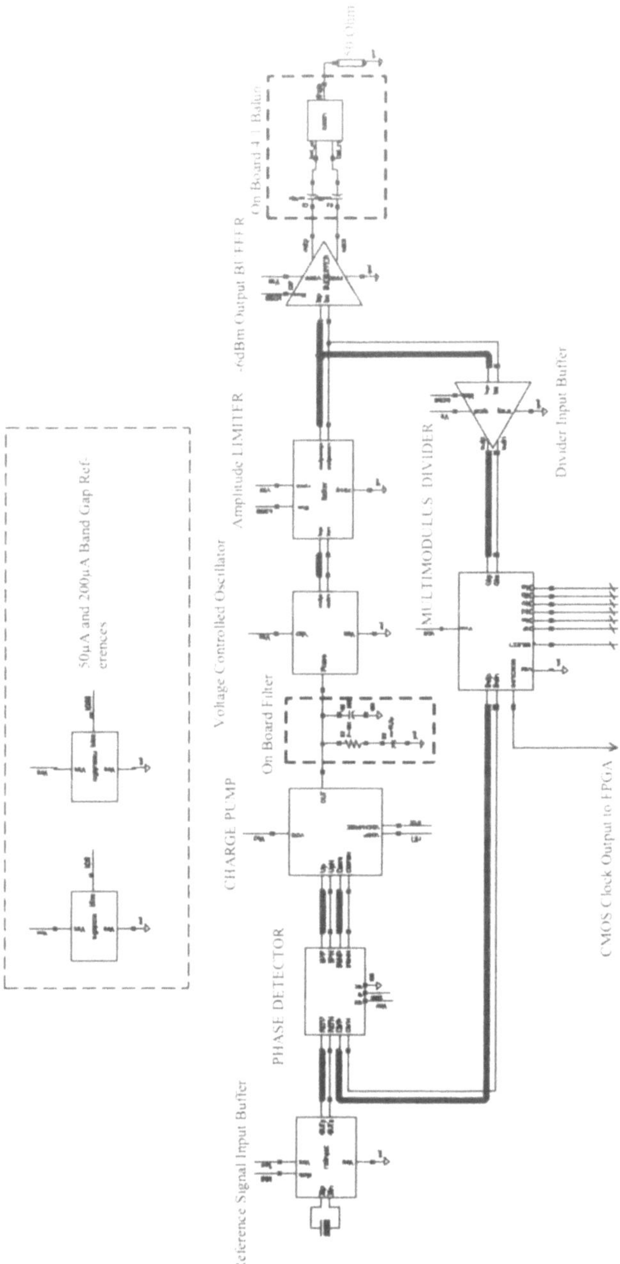

Figure 20. Complete PLL schematic

Phase Frequency Detector

The basic schematic of a phase detector circuit is shown in Figure 21.

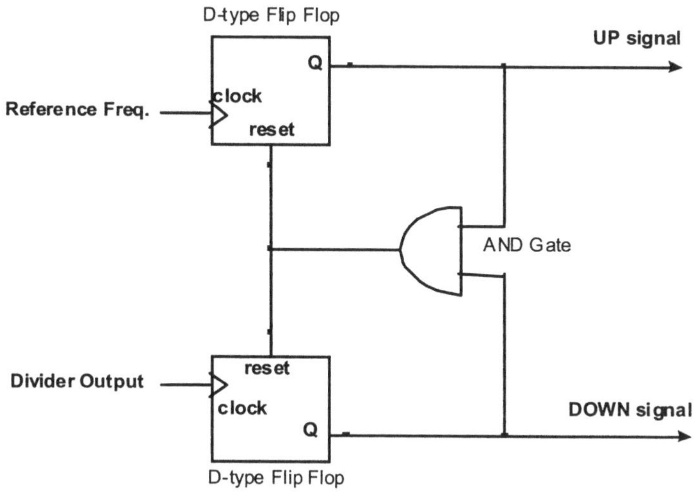

Figure 21. Phase detector structure

The circuit works as follows: if reference signal goes high before the divider output, then the UP signal goes high.

The UP signal remains high as long as the divider output signal goes high, thereafter the AND gate output goes high resetting flip flops and, in turn, lowering the UP signal.

The DOWN output remains low the whole time (except for a negligible lapse of time due to AND gate delay). Clearly, similar steps occur when divider output signal leads reference signal; in that case, DOWN output goes high and UP remains low.

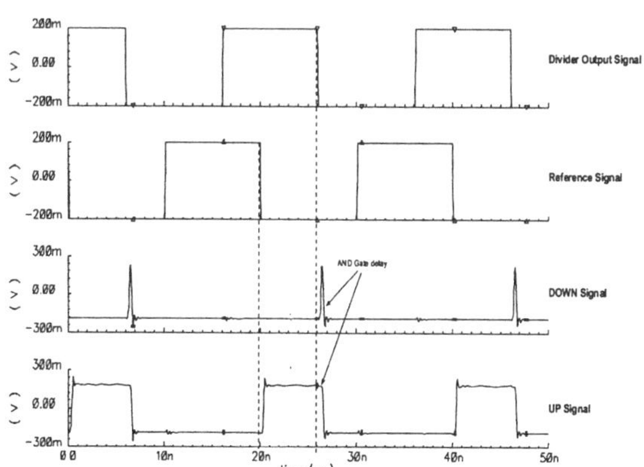

Figure 22. Phase detector transient simulation

In the PLL design, a little bit different configuration has been used. The main difference consists of the substitution of D-type flip flops with SR flip flops. Furthermore, circuit commutations happen at falling edges rather than at the leading edges of signal clock. In Figure 22, Cadence SPECTRE transient simulations of the implemented phase detector circuit are reported when a *50 MHz* reference signal is applied.

Since digital blocks are designed using differential ECL, current consumption can be traded for circuit speed. Simulations showed that if higher dissipation is allowed, phase detector still works well when 500 MHz signal inputs are applied. Despite the high frequency performance, this circuit has a main drawback for the reason that it gives a double output (UP, DOWN). Hence, the charge pump circuit has to perform a parallel to single ended conversion.

Charge Pump

The main role of the charge pump is to realize a current switch to supply the loop filter (Figure 23):

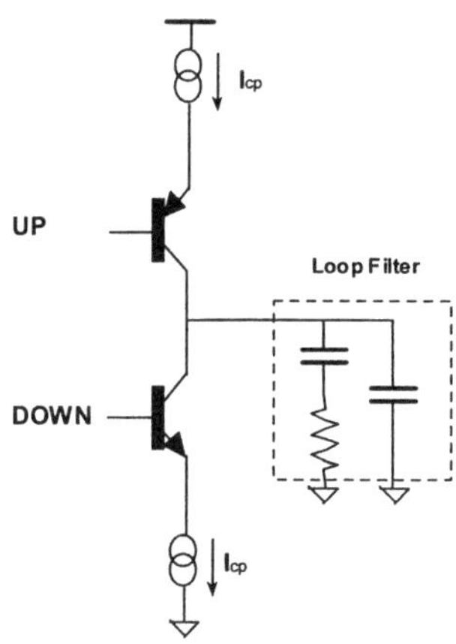

Figure 23. Charge pump basic implementation

The theoretical circuit shown above needs the adopted technology to provide a good *pnp* device. Because this is not our case, the circuit implementation uses pMOS transistors as complementary devices. Furthermore, because UP and DOWN signals come from a differential ECL phase detector, two differential to single ended buffers are needed. These buffers drive an all current mirror configuration that realizes the current switching feature (Figure 24).

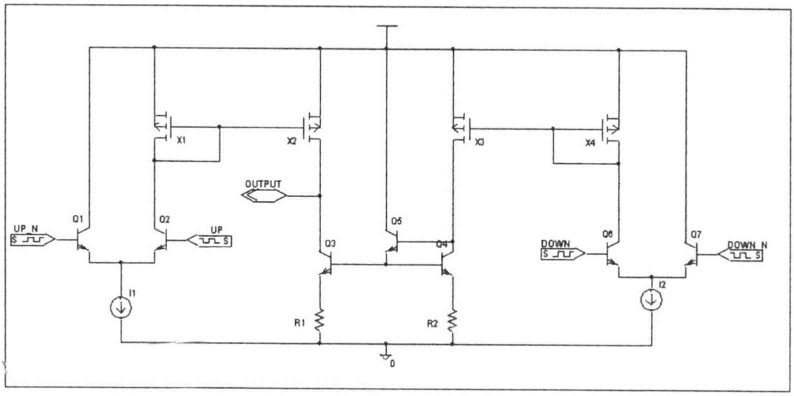

Figure 24. All current mirrors charge pump implementation

The above circuit shows a good frequency performance and allows an excellent matching of the loop filter current sources (I_{cp} in Figure 23) because I_1 and I_2 current sources of Figure 24 are generated by the same current mirror circuit. Current matching is most important because it affects the phase detector-charge pump assembly error.

In the design of a charge pump circuit for PLL applications, the choice of the current I_{cp} is a critical point. In fact, charge pump current sets the phase detector gain K_ϕ, and then directly affects the PLL transfer function (Figure 5). On the other side, as bias currents I_1 and I_2 grow up, the charge pump noise contribute becomes higher because of pMOS flicker noise. Furthermore, higher currents means a wider pMOS device area that results in a worse frequency behaviour of the circuit due to the increased strain capacitance.

The adopted choice is to provide a pin input to allow external control of the bias current. In this way, the best value for both PLL gain, charge pump noise and speed can be fixed during test measurements.

A further enhancement has been adopted to improve the behaviour of pMOS current mirros *X1-X2*, *X3-X4* (Figure 25):

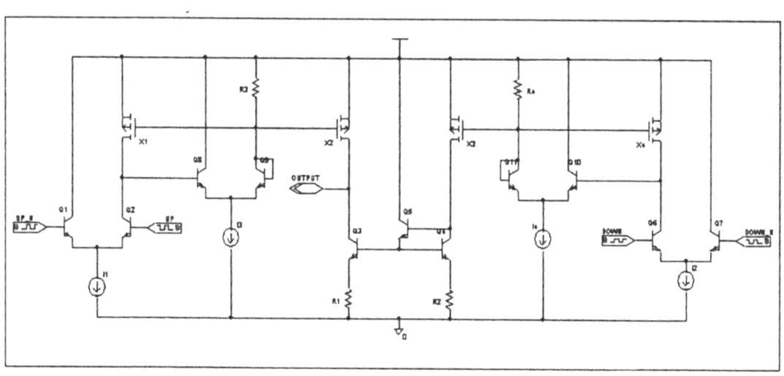

Figure 25. Enhanced charge pump schematic

Here, the pMOS diode connection is substituted with an *npn* current switching circuit.

Voltage Controlled Oscillator

A fully integrated VCO design is aimed at both a low phase noise content and a tuning range wide enough to cover target system bandwidth despite of process variations. Even though current wireless communications systems, such as UMTS, cover a few tenth of megahertz wide band of frequencies, integrated passive devices employed in resonating circuits suffer high tolerances and parasitics often making the oscillator miss desired work frequency.

Let's take into consideration a classical LC resonator negative resistance VCO (Figure 26):

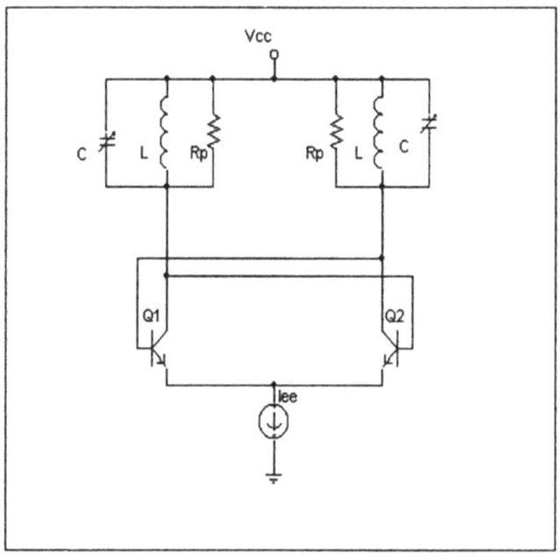

Figure 26. Basic LC resonator negative resistance oscillator

The oscillation conditions are given by:

$$\begin{cases} f_o = \dfrac{1}{2\pi\sqrt{LC}} \\ g_{m_{1,2}} \geq \dfrac{1}{Rp} = \dfrac{1}{\omega LQ} \end{cases} \tag{15}$$

where Q is the resonator quality factor. From the first of the (15), it comes that if capacitance C is modulated the oscillation frequency f_o can be tuned. In bipolar technologies, usually variable capacitances are realized using bipolar junction diodes (varactors). In fact, diode reverse bias voltage variations modulate the depletion layer width that directly sets junction capacitance.

In recent high frequency bipolar technologies, transistor base-collector junction or base-emitter junction could be used. Usually, the latter suffers a too low breakdown voltage (often less than 2 V), preventing its use as a wide tuning range varactor.

On the other side, base-collector junction shows a higher breakdown voltage ; however due to the transistor structure (Figure 27), it suffers higher parasitics.

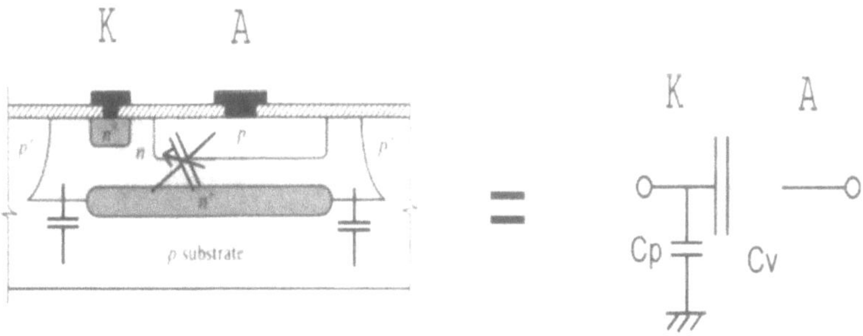

Figure 27. Varactor physical structure

As the highest strain capacitance is concentrated in the cathode terminal **K**, practical implementation of LC oscillator follows the configuration of Figure 28:

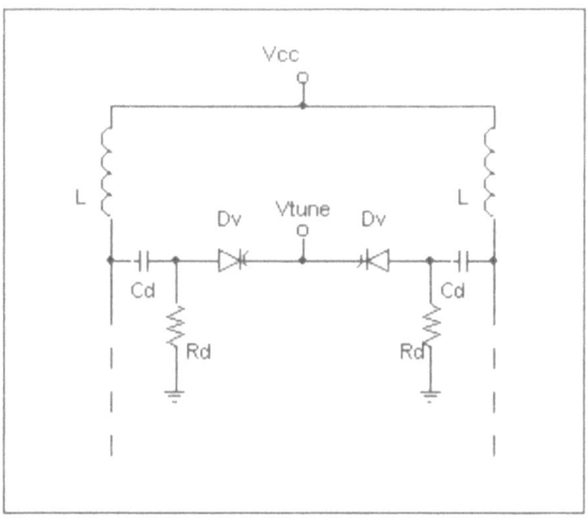

Figure 28. Varactor diodes common implementation

Here, the common cathode terminal is a signal ground and then varactor strain capacitances are not included in the LC resonator path. On the other side, to reverse bias varactors, decoupling capacitances C_d are needed. Clearly, these series devices cause a reduction of the total capacitance C_t variation an then oscillator tuning range drop. In fact, we have:

$$Ct = \frac{CvCd}{Cd + Cv} \tag{16}$$

if varactor capacitance C_v has a variation range given by:

$$\frac{\Delta Cv}{Cv_{MAX}} = 1 - \frac{Cv_{MIN}}{Cv_{MAX}} \tag{17}$$

then variation range of C_t is given by:

$$\frac{\Delta Ct}{Ct_{MAX}} = 1 - \frac{Cv_{MIN}}{Cv_{MAX}} \cdot \frac{Cv_{MAX} + Cd}{Cv_{MIN} + Cd} \leq \frac{\Delta Cv}{Cv_{MAX}} \tag{18}$$

Clearly, from (18) it comes that the tuning range reduction can be neglected as more C_d is greater than C_v. However, the size of decoupling capacitance C_t is limited by its parasitics (Figure 29).

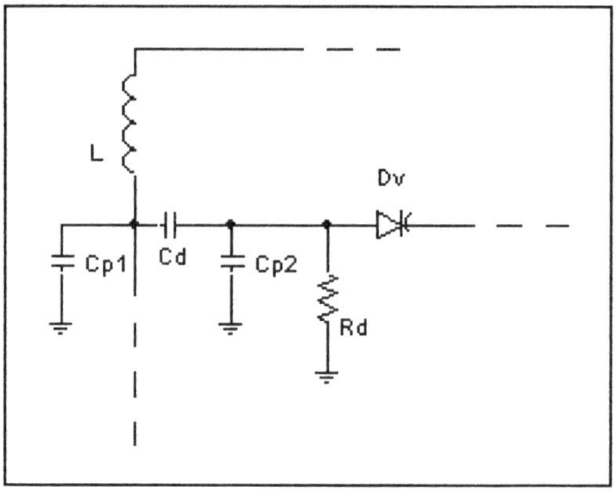

Figure 29. Decoupling capacitance parasitics

Since C_{p1} and C_{p2} may be as great as 5% of C_d, common configurations like this show a tuning range less than 250 MHz.
To overcome this limit different circuit designs has been investigated. Meissner's oscillator is showed in Figure 30, where bias details are omitted.

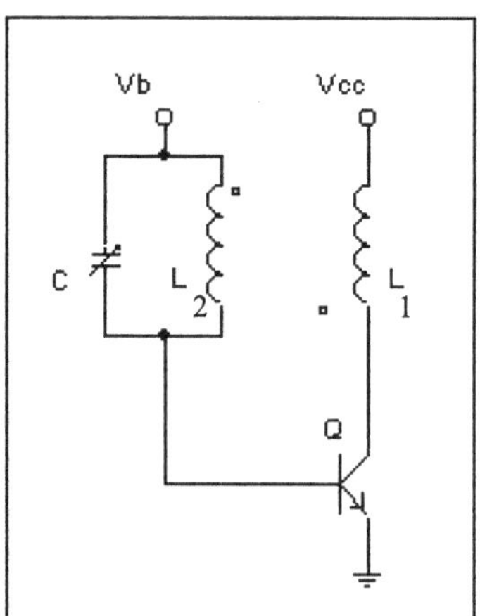

Figure 30. Meissner's oscillator

In order to derive oscillation conditions, a small signal model circuit has been considered (Figure 31):

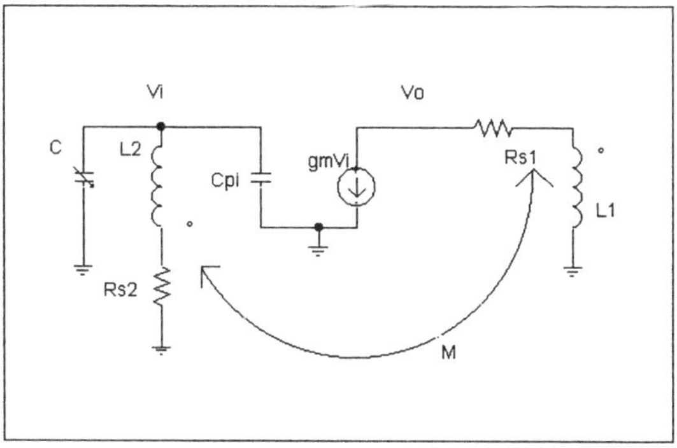

Figure 31. Oscillator small signal model

In the above schematic, R_{s1} and R_{s2} model inductor losses and M is defined as the mutual inductance between L_1 and L_2.

Since the high work frequency, transistor input resistance has been neglected compared to the reactance of input capacitance ($1/\omega C_{pi}$). In feedback circuits, oscillation occurs when forward path gain ($A=V_o/V_i$) and feedback path gain ($\beta=V_i/V_o$) fulfil Barkhausen's criterion $A\beta=1$. Therefore, if we define:

$$Z_1 = Rs_1 + j\omega L_1$$
$$Z_2 = Rs_2 + j\omega L_2 + \frac{1}{j\omega C_t} \qquad (C_t = C_{pi} + C) \tag{19}$$

we have:

$$A = \frac{Vo}{Vi} = -g_m Zo = -g_m \left[Z_1 + \frac{(\omega M)^2}{Z_2} \right] \tag{20}$$

whereas, owing to mutual coupling M, we have:

$$\beta = \frac{Vi}{Vo} = -\frac{j\omega M}{\left[Z_1 Z_2 + (\omega M)^2 \right]} \cdot \frac{1}{j\omega C_t} \tag{21}$$

and then:

$$A\beta = -g_m \left[Z_1 + \frac{(\omega M)^2}{Z_2} \right] \cdot \left\{ -\frac{j\omega M}{\left[Z_1 Z_2 + (\omega M)^2 \right]} \cdot \frac{1}{j\omega C_t} \right\} = 1 \tag{22}$$

that gives:

$$g_m \frac{j\omega M}{Z_2} \frac{1}{j\omega C_t} = 1 \tag{23}$$

substituting Z_2 :

$$g_m \frac{M}{\left[Rs_2 + j\left(\omega L_2 + \frac{1}{\omega C_t}\right)\right] C_t} = 1 \tag{24}$$

finally, oscillation conditions are:

$$\begin{cases} fo = \dfrac{1}{2\pi\sqrt{L_2 C}} \\ g_m = \dfrac{Rs_2 C}{M} \end{cases} \tag{25}$$

Defined k as the coupling factor between L_1 and L_2, if we set:

$$L_1 = L_2 = L$$
$$M = k\sqrt{L_1 L_2} = kL$$

then the minimum device transconductance that guarantees oscillation is:

$$g_m = \frac{RsC}{M} = \frac{RsC}{kL} = \frac{\omega C}{kQ} = \ldots = \frac{1}{k}\frac{1}{\omega LQ} \tag{26}$$

If we look at the second statement of the (15), since $k<1$, the minimum g_m increases compared to the classical LC negative resistance VCO. This fact means a greater current consumption and a greater shot noise sources contribution. Furthermore, in the above analysis transistor parasitics have been neglected. When collector to substrate capacitance is taken into account (Figure 32), it shows that the circuit has a double resonance structure.

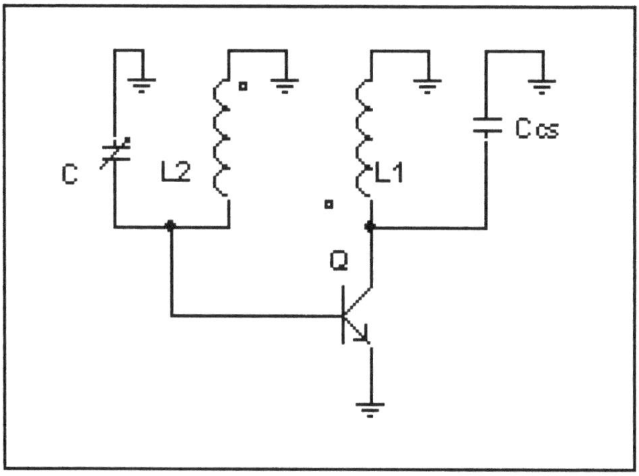

Figure 32. Parasitics in Meissner's oscillator

Therefore, the circuit is affected by a spur oscillation due to L_1 and C_{cs}. As C_{cs} is a small capacitance, spur oscillation occurs at high frequencies; so that a good design workaround is to ensure that circuit gain adequately rolls off at high frequencies. Furthermore, whenever oscillator output is taken at the L_2-C resonator terminal a likely spur rejection filtering is made.

From Leeson's approximation, oscillator phase noise can be calculated by the following expression:

$$L(\Delta f) = 10\log\left[\frac{kT}{A_0^{\ 2}}\frac{1}{Q^2}\left(\frac{fo}{\Delta f}\right)^2(1+F)\right] \tag{27}$$

where Δf is the frequency offset from the carrier f_o, A_o is the oscillation amplitude and F a factor due to active device noise contribution. In a well designed fully integrated oscillator resonator quality factor Q sets phase noise content.

In Figure 33, Meissner's configuration is applied to a cross coupled differential stage:

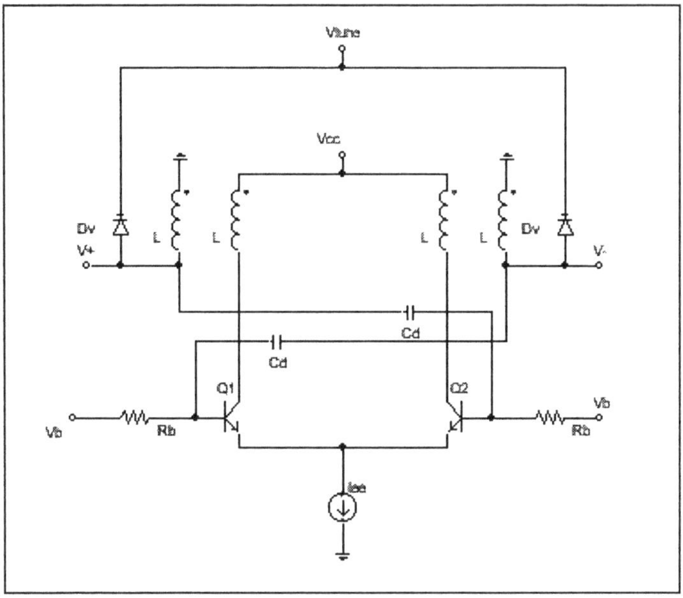

Figure 33. Cross coupled differential stage Meissner's oscillator

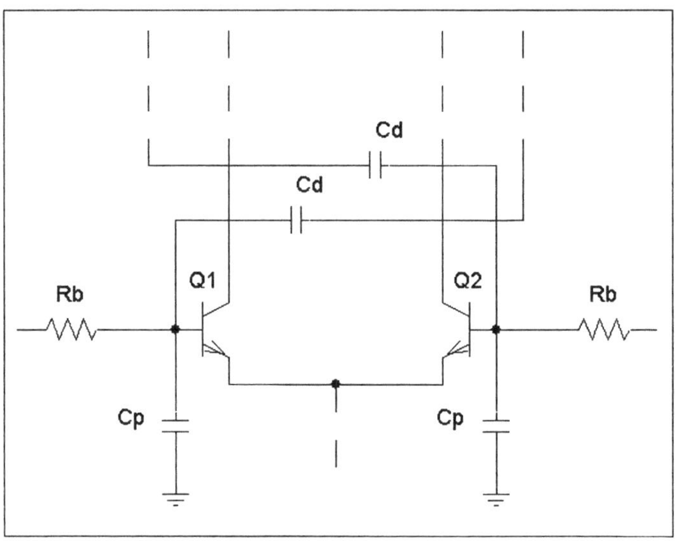

Figure 34. Decoupling capacitance connection

As shown, since varactor anodes have a zero DC bias no series decoupling capacitances are required. Capacitances C_d are required to decouple resonator from active devices bias circuit. Thus, neglecting small strain capacitances of both C_d and the integrated inductor, the whole varactor capacitance variation can be exploited.

In order to reduce base resistance noise contribution, Q_1 and Q_2 consists of four 12 μm wide emitter devices. Collector to substrate capacitances limit the use of more devices. In fact, although spur tone oscillation occurs at a high frequency, as C_{cs} increases it reduces tuning range, because mutual coupling makes it load resonator (Figure 32). Resistances Rb prevent transistor bias circuit load resonator; furthermore, they avoid that noise voltage source associated to transistor base resistances feed directly the active devices Q_1, Q_2.

Decoupling capacitances C_d are chosen with the minimum size that guarantees a low impedance path at the work frequencies. The C_d capacitances are connected in order to have the highest parasitics on the transistor base sides (Figure 34).

This choice avoids tuning range reduction and moreover helps phase noise content. In fact, from (27) it comes that a high oscillation amplitude lowers the phase noise level. In the schematic of Figure 33, oscillation amplitude is limited by base collector junction turn on occurrence. As the strain capacitance of Figure 34 makes a partition of the signal coming from resonator, it allows greater oscillation amplitude. Finally, the current source I_{ee} of Figure 33 has been implemented with a resistance rather than with an active device configuration. In this way, when either Q1 or Q2 is off, any noise contribution coming from the current source is minimized.

The fully integrated oscillator realization requires the coupled inductors to be designed. To achieve this aim, the HP Momentum CAD tool has been used. A vertical inductor coupling has been preferred over the other solutions, as it shows a greater coupling factor k.

On the other side, a drawback is that both inductors couldn't be realized with the top most and less resistive metal layer. A 2 turn circular inductor shape has been considered, the wire width was 25 μm and a 3 μm spacing. HP Momentum simulations calculate a 730 pH inductors with a quality factor about 8 for the top most metal layer inductor and about 3 for the other. Under-pass inductor connections have been taken into account for the simulation. Furthermore, the simulated coupling factor is $k=0.83$. Simulation plots are reported in Figure 35 and Figure 36:

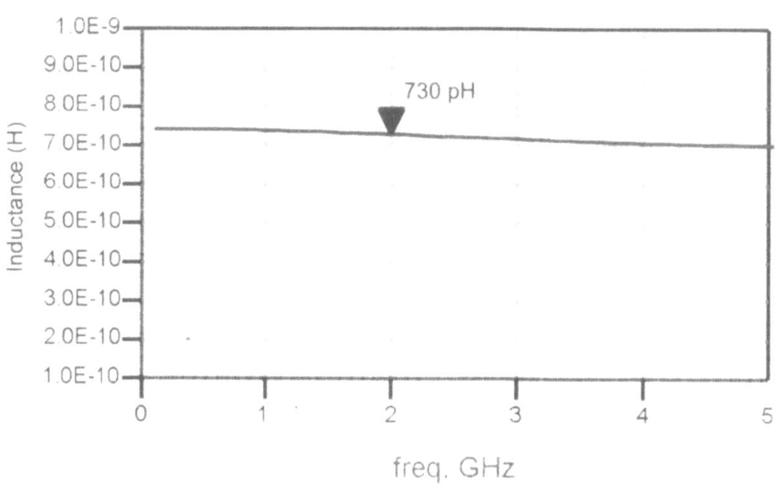

Figure 35. Simulated inductance value

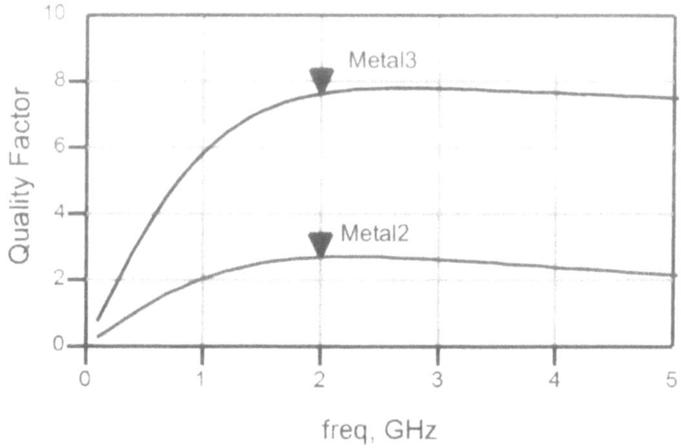

Figure 36. Simulated inductor quality factor Q

The VCO has been studied with both Cadence SPECTRE and HP ADS simulators. Phase noise level is about –104 dBc/Hz at 100 kHz offset from the carrier. The theoretical tuning range results greater than 600 MHz, whereas tuning slope *Kv* is about 300 MHz/V with a 2 GHz oscillation frequency. Furthermore, second harmonic tone power results about 35dB lower than fundamental. In the following figures, main significant simulation plots are reported.

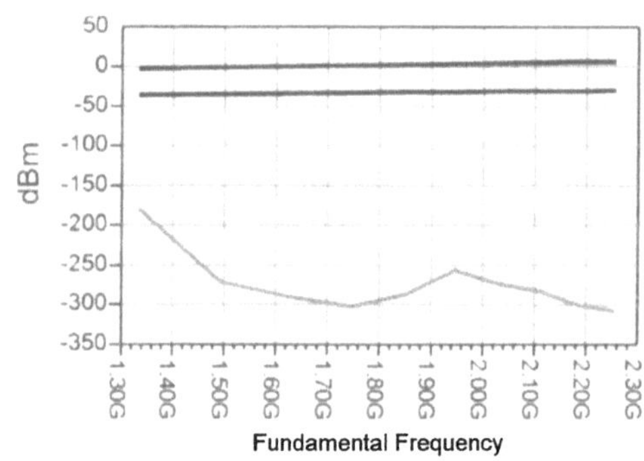

Figure 37. Fundamental and harmonics vs frequency

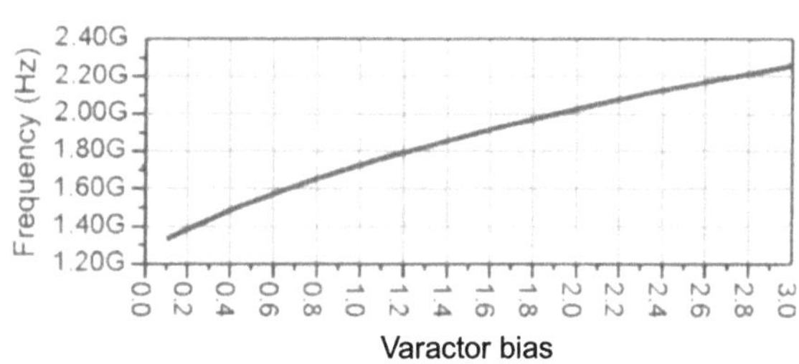

Figure 38. Fundamental frequency vs varactor bias

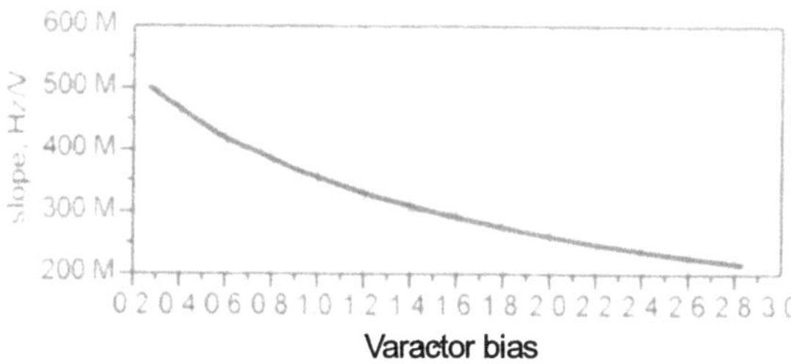

Figure 39. Tuning slope vs varactor bias

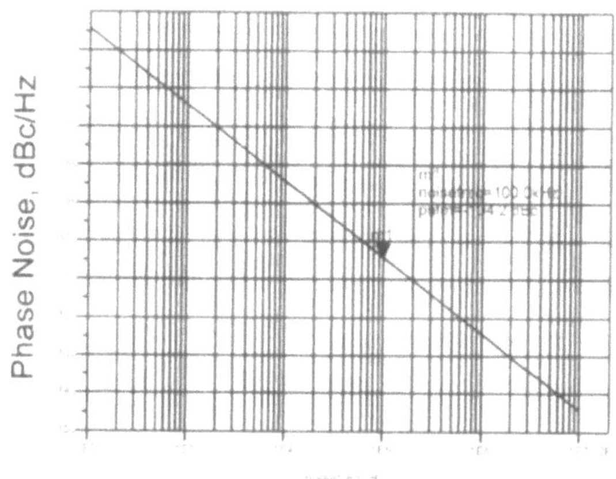

Figure 40. Oscillator phase noise

Practical implementation of the VCO has to provide an amplitude control circuit. In fact, from Leeson's equation comes that phase noise depends strictly on the oscillation amplitude.

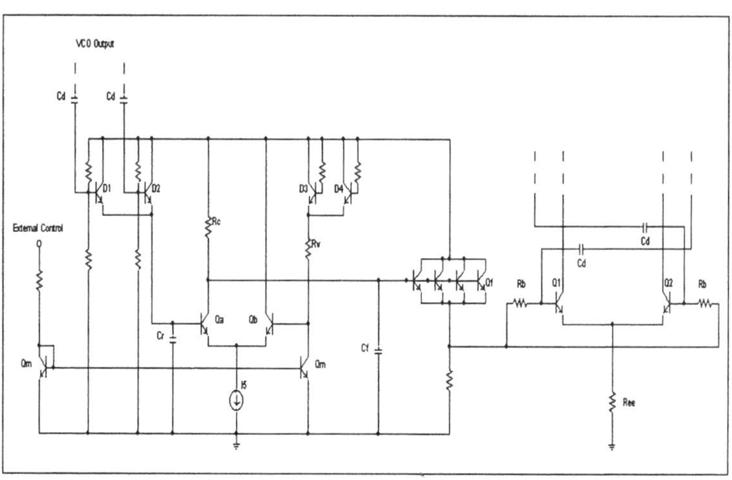

Figure 41. Amplitude control circuit

Because of both passive device tolerances and parasitics, wanted oscillation amplitude could be missed. On the other side, if too high an amplitude occurs then the-circuit may not work properly, because of device saturation. In this work an amplitude control circuit has been designed (Figure 41).

The above circuit works as follows: VCO output is rectified and filtered by D_1, D_2 and C_r. The differential stage Q_a, Q_b compares voltage drop over C_r to a reference voltage externally set by the current mirror Q_m.
Finally, amplified error voltage regulates bias of VCO active devices Q_1, Q_2 through Q_f. In order to reduce base resistance noise contribution, Q_f consists of four 12 μm wide devices. Clearly, capacitance C_f guarantees overall loop stability.

In Figure 42, transient simulations of the oscillator with its amplitude control circuit are reported.

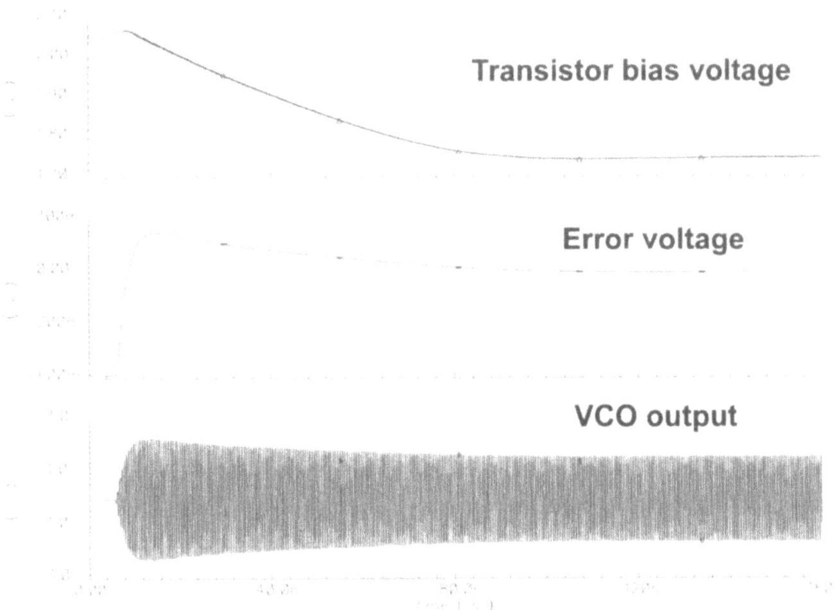

Figure 42. VCO with amplitude control circuit simulations

Multi-modulus Divider

Programmable divider basic structure consists of a 2/3 dual modulus divider cascade (Figure 43).

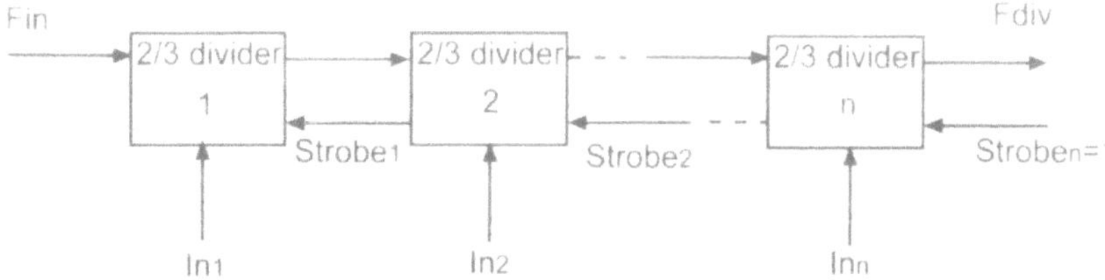

Figure 43. Divider basic structure

The inputs In_i enable the 2/3 cells to divide by 3 when the $Strobe_i$ signal is high. Furthermore, $Strobe_i$ signal propagates from n^{th} to first cell and, for each 2/3 block, it goes high only once in a division cycle. Operating in that way, the circuit allows to divide input frequency in the range $[2^n, 2^{n+1}-1]$. Because work frequency of 2/3 cells decreases throughout the chain, current consumption of each block can be designed for optimal performance.

The schematic of a 2/3 divider cell is shown in Figure 44.

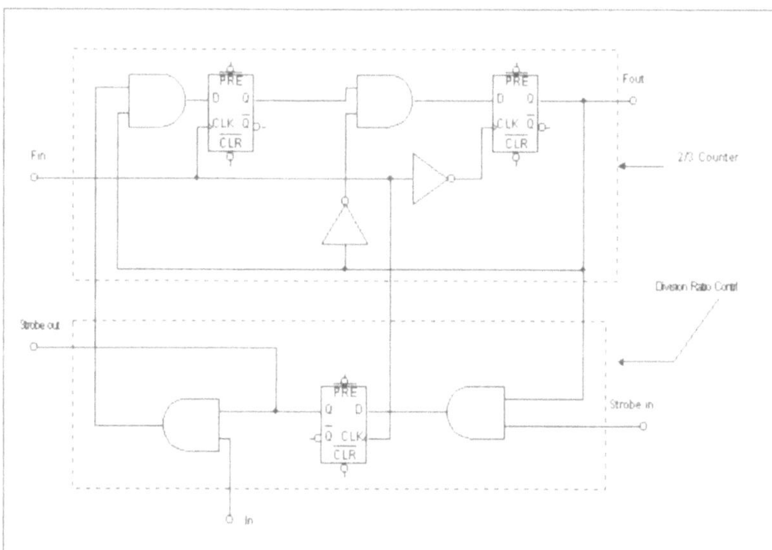

Figure 44. 2/3 divider schematic

As shown, the cell consists of a *D*-latch 2/3 counter and a digital network that decodes division ratio, given *In* and *Strobe* inputs. The circuit has been designed for properly work at 2.5 GHz as Cadence SPECTRE simulation shows in Figure 45.

Figure 45. 2/3 divider simulation

In the simulation, the 2/3 cell has been set for divide by 3 operation. In order to use a PLL reference frequency comprised in a range that goes from 20 MHz to 50 MHz, if PLL output frequency is around 2 GHz, needed division ratios go from *32* to *127*. The whole range has been split in two ranges: [32, 63] and [64, 127]. To cover the latter, six 2/3 divider cells are employed. Whenever a division ratio lower than *64* is required, a dedicated logic circuit disables the 6th cell of the chain and outputs the 5th cell signal as F_{div}.

Furthermore, CMOS compatible interfaces have been provided, in order to make programmable divider be controlled by ΣΔ modulator implemented on FPGA. In Figure 46, the complete multi-modulus divider schematic is reported whereas in Figure 47 transient simulation results are displayed when the circuit is set to divide by 65.

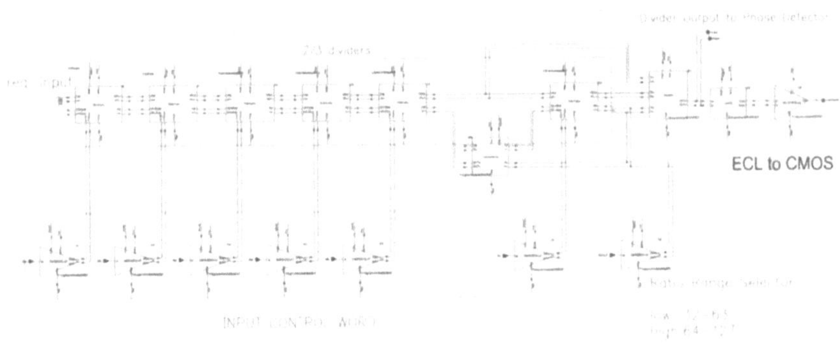

Figure 46. Multi-modulus divider, complete schematic

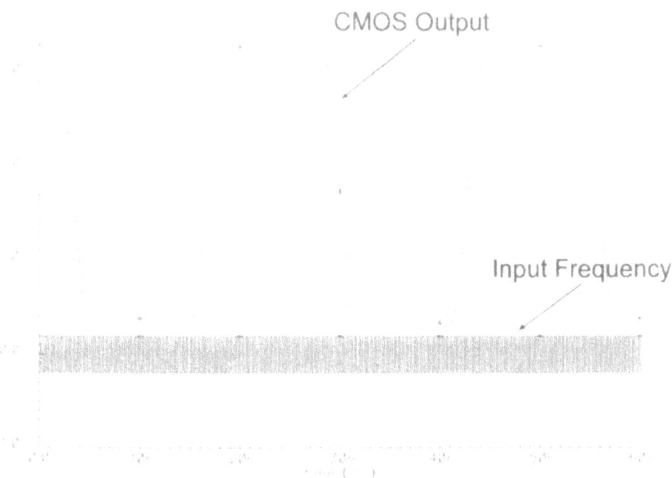

Figure 47. Multi-modulus divider transient simulation

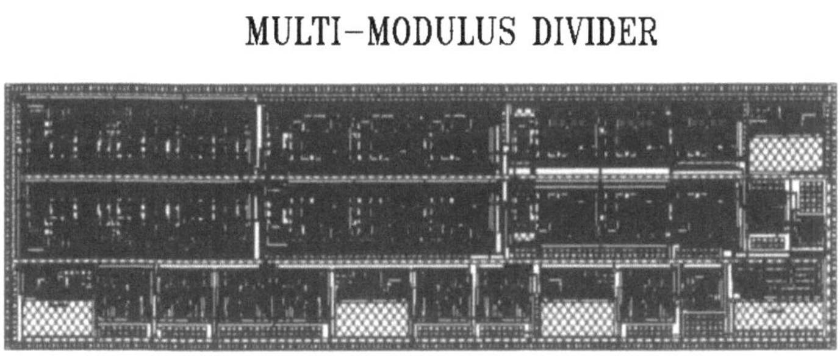

Figure 48. Programmable divider layout

In Figure 48, the layout view of multi-modulus divider has been reported. Current consumption is 28 mA when a 3.3 V source is supplied.

Final Remarks

Because of PLL design complexity, any other analysis but transient simulation of the whole circuit is not practical. Transient simulations have been run in order to ensure PLL system correct operation whereas overall noise performance has been extrapolated by single block noise investigations.

Using PLL linear approximation of Figure 5, the overall phase noise level resulted about –110 dBc/Hz @100 kHz achieving design goal (Figure 48). We have to observe that simulations didn't take into account flicker noise contribution, because related device model parameters have not been provided for the adopted technology. Finally, the circuit draws about 70 mA from a 3.3 V supply.

In Table 1, the PLL parameters summary is shown:

Table 1. PLL parameters summary

Reference Frequency	60 MHz
Loop Bandwidth	500 kHz
Output Frequency	1.98 GHz
Phase Noise @ 100 KHz	~-110 dBc/Hz

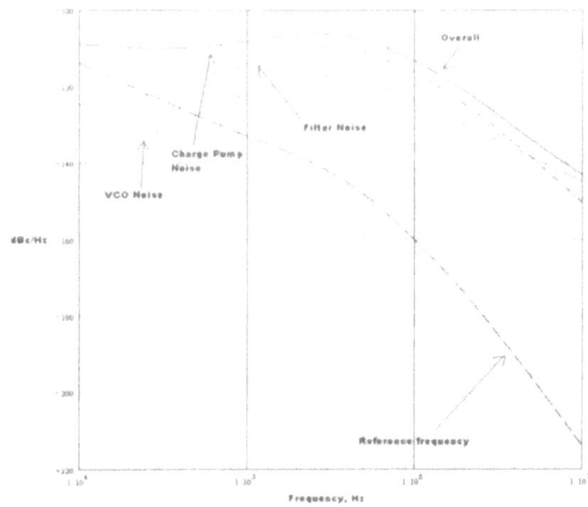

Figure 49. PLL phase noise contributors and overall performance

Conclusions

In this work a fractional-N PLL frequency synthesizer for wideband wireless application has been designed.

Because of theoretical superior performance, ΣΔ modulation has been preferred to modulate division ratio of the multi-modulus divider. This choice should guarantee both a lower phase noise content and spur-free circuit feature. Furthermore, ΣΔ modulator enables high step resolution without the use of low reference frequencies.

In order to test different structures, it was preferred to implement ΣΔ modulator on FPGA. The remaining PLL blocks have been designed for high frequency bipolar technology integration, using status of the art techniques. The STMicroelectronics BiPMOS technology has been used; it provides a bipolar transistor with f_t>40 GHz and *p*-channel MOS transistor.

Because of lot-to-lot process variations, fully integrated voltage controlled oscillators have to provide a tuning range wide enough to reliably cover the target system bandwidth. In this work, in order to improve VCO tuning range, a novel circuit structure has been studied.

Finally, circuit simulations have reported that PLL phase noise level is about –110 dBc/Hz at 100 kHz offset from the carrier frequency, achieving recent wideband wireless communication systems specification.

References

[1] B. Miller and B. Conley. A multiple modulator fractional divider. IEEE Transactions On Instrumentation and Meas, 40 (3), June 1991.

[2] V. Friedman. The structure of the limit Cycles in Sigma Delta Modulation. IEEE Transactions On Communications, 36 (8), August 1988.

[3] T.A.D. Riley, M.A. Copeland, T.A. Kwasniewski. Delta-sigma modulation in fractional-N frequency synthesis. IEEE J. Of Solid-State Circuits, 28, May 1993.

[4] H. Wang. A Geometric View of ΣΔ Modulations. IEEE Trans. On Circuits and Systems-II Analog and Digital Signal Processing, 39 (6), June 1992.

[5] C. Vaucher and D. Kasperkovitz. A Wide-Band Tuning System for Fully Integrated Satellite Receivers. IEEE J. Of Solid-State Circuits, 33 (7), July 1998.

[6] N.H. Sheng, R.L. Pierson, K.C. Wang, R.B. Nubling, P.M. Asbeck, M.C.F. Chang, K.L. Edwards, and D.E. Phillips. A High-Speed Multimodulus HBT Prescaler for Frequency Synthesizer Applications. IEEE J. Of Solid-State Circuits, 26 (10), October 1991.

[7] C. Vaucher et al. A Family of Low-Power Truly Modular Programmable Dividers in Standard 0.35-μm CMOS Technology. IEEE J. Of Solid-State Circuits, 35 (7), July 2000.

[8] T.H. Lee. Oscillator Phase Noise: A Tutorial. IEEE J. Of Solid-State Circuits, 35 (3), March 2000.

[9] M.A. Margarit et al. A Low-Noise, Low-Power VCO with Automatic Amplitude Control for Wireless Applications. IEEE J. Of Solid-State Circuits, 34 (6), June 2000.

[10] J. Craninckx and M.S.J. Stejaert. A 1.8-GHz Low-Phase-Noise CMOS VCO Using Optimizied Hollow Spiral Inductors. IEEE J. Of Solid-State Circuits, 32 (5), May 1997.

[11] M. Zannoth, B. Kolb and J. Fenk. A Fully Integrated VCO at 2 GHz. IEEE J. Of Solid-State Circuits, 33 (12), December 1998.

Analysis and Improvement of a Multi-channel Amplifier Integrated Circuit for Rad-hard Applications

Salvatore G. Pastorina

STMicroelectronics S.r.l.
Stradale Primosole 50, 95100 Catania – Italy

e-mail: salvatore.pastorina@st.com

Introduction

The subject of this paper is the analysis and the improvement of a multi-channel amplifier integrated circuit first designed in the Italian INFN (National Institute of Nuclear Physics) laboratories and than manufactured by ST Microelectronics.
Since the circuit has never worked properly, several measurements have been carried out but no evidence of the origin of the problem has been found. In the following chapters first we will describe the structure and the behaviour both of the whole chip and of its single channels, then we will try to explain why the experimental response is so different from the simulated one. Finally we will propose an improved version of the circuit to be implemented through simply bridging some stages of the original one.

The 16 Channels INFN Chip

Although our circuit is basically an amplifier, it differs from a common one in two aspects:

a) it should be able to track the impulse signal coming from a charge detector and hold its peak value throughout a time interval set by a trigger signal;
b) its destination on a satellite makes it a rad-hard class circuit.

As a consequence of point b) it is not possible for us to use whatever design technique (*i.e.*, we can't do a wide use of MOSfets).

Figure 1 shows the structure of the actual device. It is made of 16 independent channels and a test one; their outputs are all connected to a common bus, each through a multiplexer controlled transmission gate. In this way only one channel at a time can communicate with the chip output buffer.

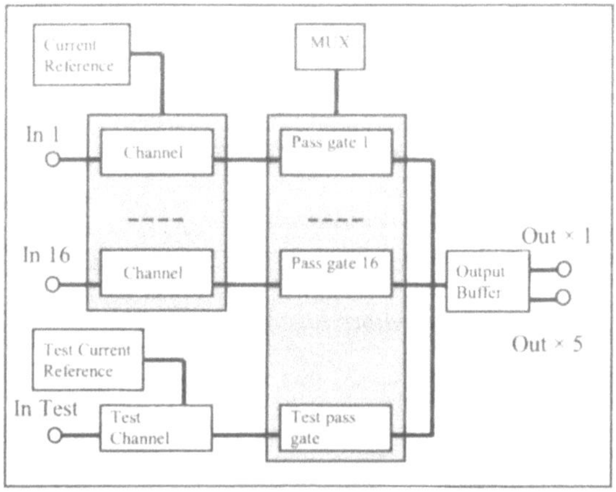

Figure 1. INFN chip block diagram

In each channel we can find three different functional blocks (see Figure 2).

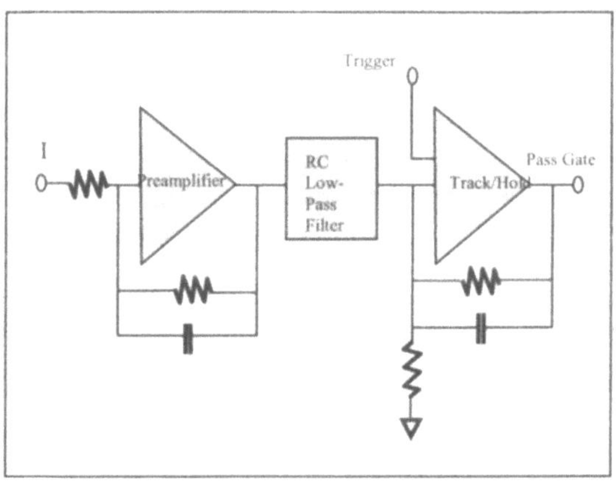

Figure 2. Single channel block diagram

The first one is a transresistance amplifier which converts the current signal, coming from the charge detector, into a voltage one.

The latter, then, can be elaborated by the subsequent stages. Next to the preamplifier we can find a RC low pass filter; it is used for giving the signal a gaussian like shape.

Finally, the last functional block is a track/hold circuit. How we said previously, this circuit detects and holds the peak value of its input signal; it also provides some gain to compensate for the attenuation introduced by the preamplifier and the filter.

Signal Flow through the Preamplifier and the RC Shaping Filter

We shall now illustrate the circuit used to implement the transresistance amplifier and the low pass filter. For this purpose see Figure 3. The amplifier consists basically of a cascade pair followed by a common collector, some biasing and a shunt-shunt feedback network; the filter is a very simple RC circuit.

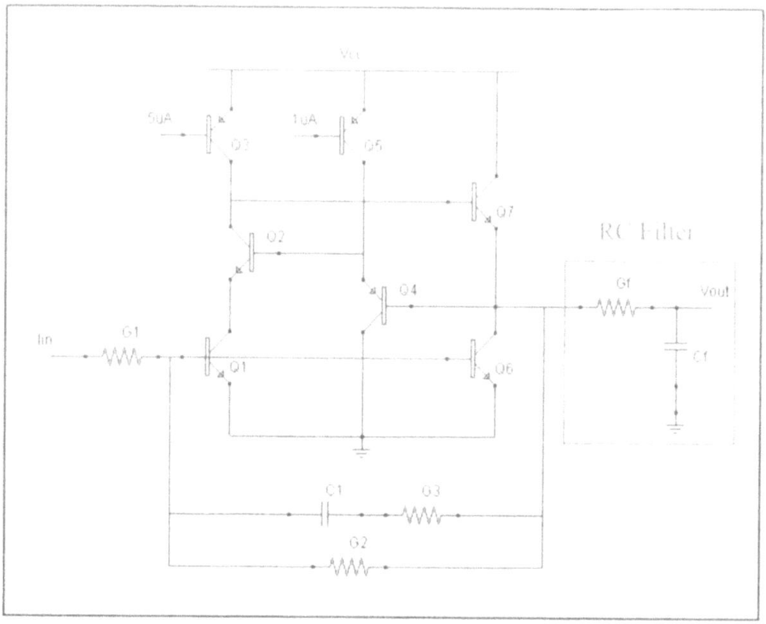

Figure 3. Input transresistance preamplifier and RC shaping circuit

We are interested in the input signal changes when it flows throughout these two blocks.

Let's consider the small signal equivalent circuit of Figure 4:

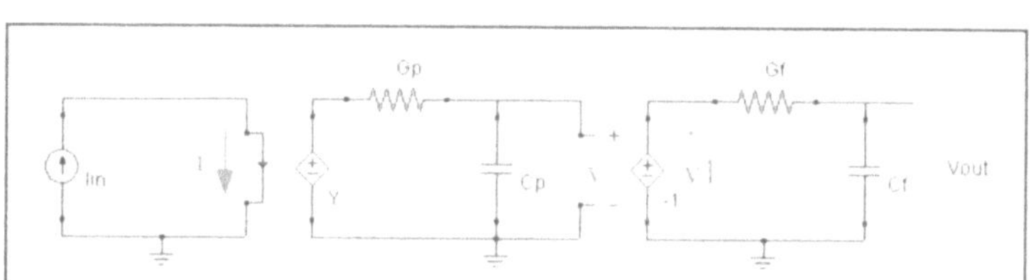

Figure 4. Small signal equivalent circuit of the input preamplifier and the RC filter where the RC network G_p, C_p models the finite bandwidth of the preamplifier circuit ($C_p/G_p=1/2\pi BW$[1]).

[1] Preamplifier bandwidth

We have:

$$I_{in}(t) = -\delta(t)i_0 = \lim_{\tau_i \to 0}(-i_0)\frac{\tau_0}{\tau_i}e^{-t/\tau_i} = \lim_{\tau_i \to 0} i_{\tau_i}(t);$$

$$i_{\tau_i}(s) = -\frac{\tau_0}{\tau_i}\frac{i_0}{\frac{1}{\tau_i}+s};$$

$$v_{\tau_i} = -\frac{i_{\tau_i}(s)}{y}\frac{1}{1+R_pC_ps} = -\frac{v_0\frac{\tau_0}{\tau_i}}{1+R_pC_ps}\frac{1}{\frac{1}{\tau_i}+s} = -\frac{v_0\frac{\tau_0}{\tau_i}}{1+\tau_ps}\frac{1}{\frac{1}{\tau_i}+s},$$

where $v_o = \frac{i_o}{y}, \tau_p = R_pC_p$;

$$V_{out,\tau_i}(s) = \frac{v_0\frac{\tau_0}{\tau_i}}{1+\tau_ps}\frac{1}{1+R_fC_fs}\frac{1}{\frac{1}{\tau_i}+s} = \frac{v_0\frac{\tau_0}{\tau_i}}{1+\tau_ps}\frac{1}{1+\tau_fs}\frac{1}{\frac{1}{\tau_i}+s}, \quad \tau_f = R_fC_f;$$

and then

$$V_{out,\tau_i} = \frac{v_0}{1-\frac{\tau_f}{\tau_p}}\frac{\frac{\tau_0}{\tau_i}}{\frac{1}{\tau_i}-\frac{1}{\tau_p}}\frac{1}{\tau_p}e^{-t/\tau_p} + \frac{v_0}{1-\frac{\tau_p}{\tau_f}}\frac{\frac{\tau_0}{\tau_i}}{\frac{1}{\tau_i}-\frac{1}{\tau_f}}\frac{1}{\tau_f}e^{-t/\tau_f}$$

$$+\frac{v_0}{1-\frac{\tau_p}{\tau_i}}\frac{\frac{\tau_0}{\tau_i}}{1-\frac{\tau_f}{\tau_i}}e^{-t/\tau_i}$$

finally, for $\tau_i \to 0$

$$V_{out} = v_0\frac{\tau_0}{\tau_p-\tau_f}e^{-t/\tau_p} + v_0\frac{\tau_0}{\tau_f-\tau_p}e^{-t/\tau_f} = v_0\frac{\tau_0}{\tau_p-\tau_f}(e^{-t/\tau_p} - e^{-t/\tau_f})$$

The maximum value of V_{out} occurs when $t=t^*$ where t^* is given by

$$t^* = \frac{\tau_p\tau_f}{\tau_p-\tau_f}\ln\frac{\tau_p}{\tau_f}.$$

This result makes it possible for us to set an upper limit to the time instant the trigger signal should be provided to the circuit to activate the hold mode.

The Current Track/Hold Circuit

The circuit used to implement the track/hold block is shown in Figure 5; there are five different stages in it: an input differential pair (Q_1, Q_2), a shunt gain stage, a common emitter (Q_3), a large capacitor for peak voltage storing (C_s) and some compensating circuitry, an output buffer (Q_5, Q_6).

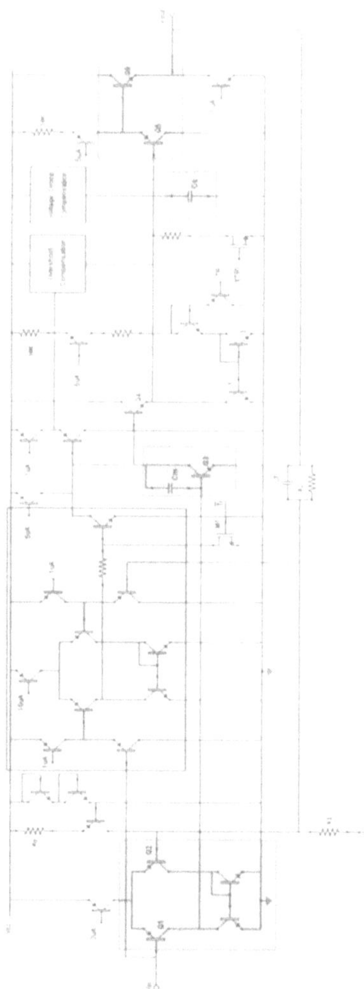

Figure 5. The track/hold circuit currently in use. In light grey from left to right: the input differential pair (Q1, Q2), the shunt gain stage, the emitter follower (Q3), the holding capacitor (Cs) and finally the output stage (Q5, Q6). F(TR) is active for a brief time interval when TR switches from hi to low.

The overall gain is determined by a series-shunt feedback network, which also provides some frequency compensation (C_f) together with the Miller capacitor Cm. When the trigger is on the peak detection is performed turning off the bias circuit of the emitter follower Q_4; when it is off the whole circuit behaves as a common amplifier. What follows illustrates the problems affecting this circuit.

AC Analysis

It is straightforward to see that the circuit will exhibit a non-dominant-pole frequency response. The responsible for the first pole is the Miller capacitor C_m:

$$p1 \cong \frac{1}{r_{\pi 3} A_{v3} C_m} = \frac{I_c}{\beta_n V_a C_m} \cong 400Hz;$$

the second one origins from Cs: the impedance seen by Cs is:

$$Zs = \frac{1}{g_m} + \frac{r_{B5}}{\beta_n + 1}$$

where r_{B5} is the total resistance at B_5, thus

$$p2 \cong \frac{1}{Z_s C_s} \cong 1.4 \; 10^5 \, Hz$$

The other poles fall beyond f_T. The simulated open-loop frequency response shown in Figure 6 confirms those results: we can see that, because of Cs, the bandwidth reduces and the phase margin is very low (it is positive thanks to the zero in the feedback path). This problem affected the first release of the circuit causing it to oscillate (it was sufficient a little capacitive load to make the circuit unstable).

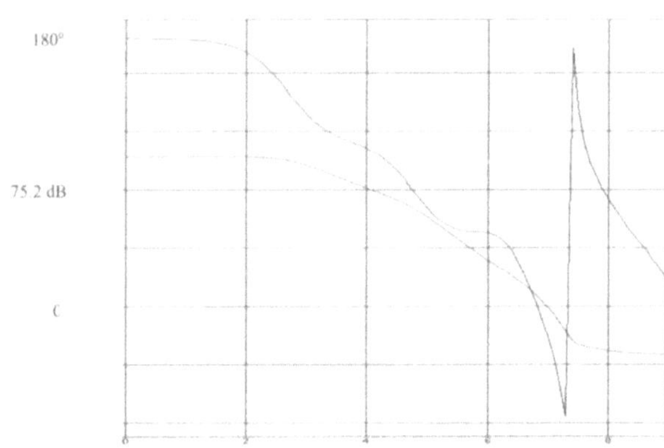

Figure 6. Frequency response without pole-zero cancellation

To overcome this problem, in the subsequent release a resistor R_s (10Kohm value) was introduced in series with C_s to compensate the pole p_2 with a zero ($z_2 \approx 1/R_sC_s$). Figure 7 illustrates the result of this compensation.

Figure 7. Frequency response with pole-zero cancellation

The net effect is an increase of the bandwidth and of the phase margin. Although the circuit is now perfectly stable, its stability margins are affected by the values of the resistors R_o and R_l which control the current in the feedback path and consequently in the emitter follower Q6. Because of the usual process tolerances on resistors (±15%), if R_o were a 15% lower than its nominal value and R_l a 15% higher, the emitter current of Q_6 and so its g_m would lower too and E_6 would become a high impedance node. Then with only a 1pF capacitor at the circuit output (it models the capacitive effects introduced by the bus and by the subsequent buffer) the pole initially originated only by C_f would move to lower frequencies and the phase margin would decrease (see Figure 8).

Figure 8. Frequency response with pole-zero cancellation; tolerances on resistors and capacitive load are considered too

Transient Requirements and Compensation Circuitry

As we said previously, the track/hold circuit is supposed both to amplify its input signal while introducing as little distortion as possible and to detect and maintain the peak value of the same signal whenever the trigger input is high. In this case the desired waveform is shown in Figure 9:

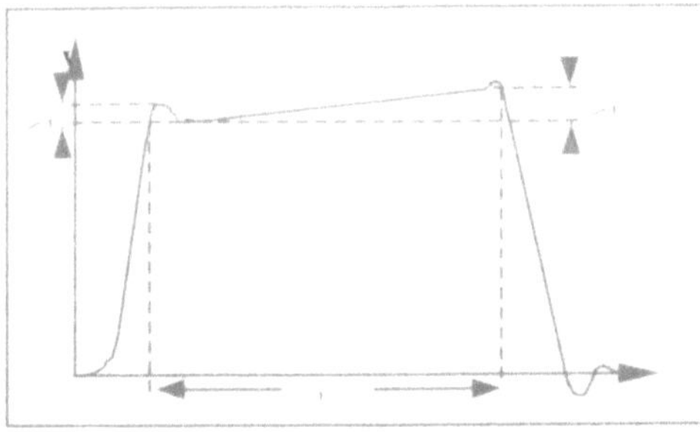

Figure 9. Transient requirements for the track/hold circuit output response (hold mode)

To meet the requirements of no more than 1mV overshoot and 1mV/100μs voltage droop across the holding capacitor C_s during the hold mode, the circuit has been provided with two compensating sub-circuits.

The first one (Figure 10) tries to compensate for the voltage overshoot caused by the base-emitter capacitance of the common collector Q4 when it turns off. To achieve this purpose the base signal of Q4 is inverted and used to drive the series of the b-e junction capacitances of two diode-connected npn transistors (one reverse and the other one forward biased).

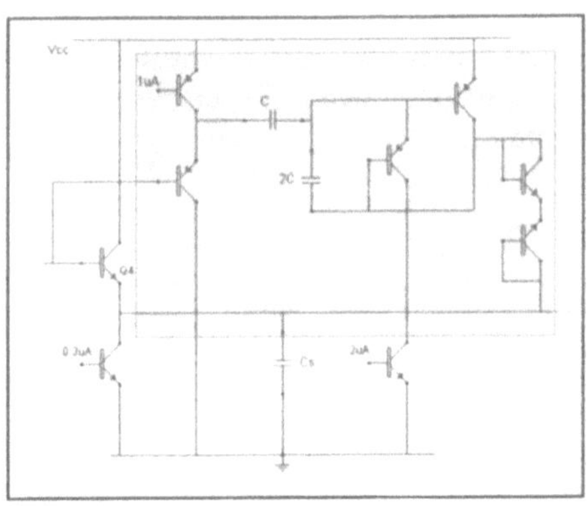

Figure 10. Overshoot compensation circuit

The second circuit (Figure 11) uses mirroring techniques to compensate for the base current of Q5 that is responsible for the excess voltage droop across C_S.

The same method is usually adopted to reduce the input currents of operational amplifiers; nevertheless there is a reason why it can't work properly here. If we want Cs not to increase/decrease its charge more than Cs 1mV in a 100μs time interval we have to limit the leakage current flowing on it below a critical value i_c given by

$$i_c = C_s \frac{\Delta V}{\Delta t} = 30pF \frac{1mV}{100\mu s} = 0.3nA$$

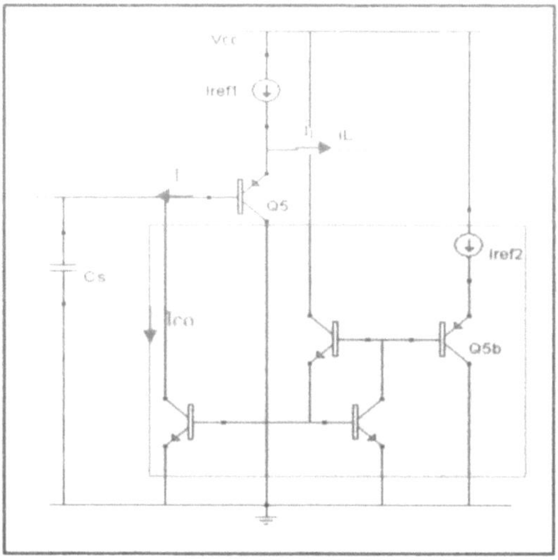

Figure 11. Voltage droop compensation circuit

This means that if I is the base current of Q_5, to meet the specifications the compensating circuit should sink a current

$$I_{co} = I \pm \Delta I, \quad -0.3nA < \Delta I < 0.3nA\,,$$

and this is a very restrictive condition. If I_{ref1}, I_{ref2}, Q_5, Q_{5b} were perfectly matched and i_L were considered too (it is not in the actual circuit) then the idea would work; in fact a theoretical analysis would lead to

$$Ico = \frac{I}{1+\dfrac{2}{\beta^2+\beta}} \Rightarrow I - I_{co} = I(1-\frac{1}{1+\dfrac{2}{\beta^2+\beta}})$$

If we now set $I-I_{co}<i_c$ and solve for β we obtain

$$\beta > \beta_{min} = \frac{1}{4}(2\sqrt{\frac{8I}{i_c}-7}-1)$$

So if I=11nA then β_{min}=8.2 and from Figure 12 we can see that even after strong irradiation this condition can be satisfied but then we have to consider that I_{ref1}, I_{ref2}, Q_5, Q_{5b} are not perfectly matched: at this current levels there is usually a pre-irradiation tolerance on βs of more than 20%; in this case (not the worst), neglecting i_L and supposed now ideal the current mirror and $I_{ref1}=I_{ref2}=I_{ref}$, we have:

$$I = \frac{I_{ref}}{\beta_5}, \quad Ico = \frac{I_{ref}}{\beta_{5B}} = \frac{I_{ref}}{\beta_5 + \Delta\beta}$$

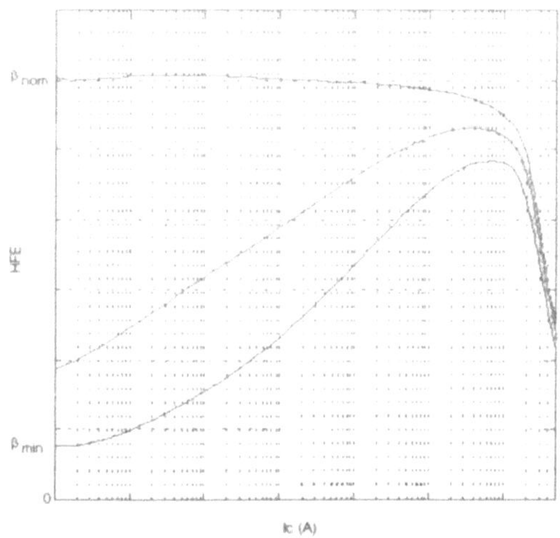

Figure 12. Pre and after irradiation qualitative β(Ic) curves

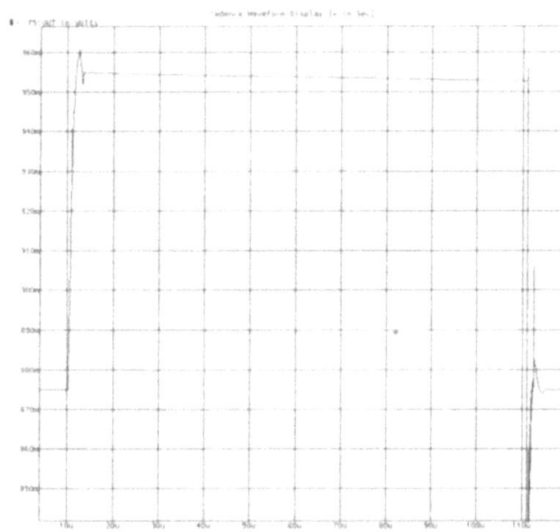

Figure 13. Current track/hold circuit output response (hold mode)

$$I - I_{co} = \frac{I_{ref}}{\beta_5} - \frac{I_{ref}}{\beta_5 + \Delta\beta} = I_{ref} \frac{\Delta\beta}{\beta_5(\beta_5 + \Delta\beta)},$$

then

$$I - I_{co} < ic \quad \Rightarrow \quad I_{ref} \frac{\Delta\beta}{\beta_5(\beta_5 + \Delta\beta)} < i_c \quad \Rightarrow \quad \Delta\beta < \frac{\frac{i_c}{I_{ref}}\beta^2}{1 - \frac{i_c}{I_{ref}}\beta};$$

with $I_{ref}=1\mu A$ and $\beta=80$ then $\Delta\beta_{max}=1.97$ or $\Delta\beta_{max,\%}=2.46\%$; so it should be evident now that the strict compensation needed to meet the desired requirement is an unachievable goal (see Figure 13).

Experimental Results

In this section we will describe and try to explain the results of the measurements performed on the chip. For this purpose refer to Figure 14.

To test the chip a voltage step was coupled via a 100nF capacitor to the input node of the track/hold circuit (trigger low: track mode). The resulting signal exhibited a very large overshoot that was not observed during the simulations. Now, the output node of the track/hold was loaded with the probe of both an oscilloscope and a multi-meter so we wandered if the insertion of such instrumentation could be the origin of the discrepancy between the simulated response and the measured one.

The idea was then to provide the schematic with a generic capacitor C_{par} modelling the effects of the external probes and then run a parametric analysis. The simulation results fitted exactly the experimental ones if $C_{par}=280pF$, so our idea was probably correct. Furthermore it was confirmed by two other facts:

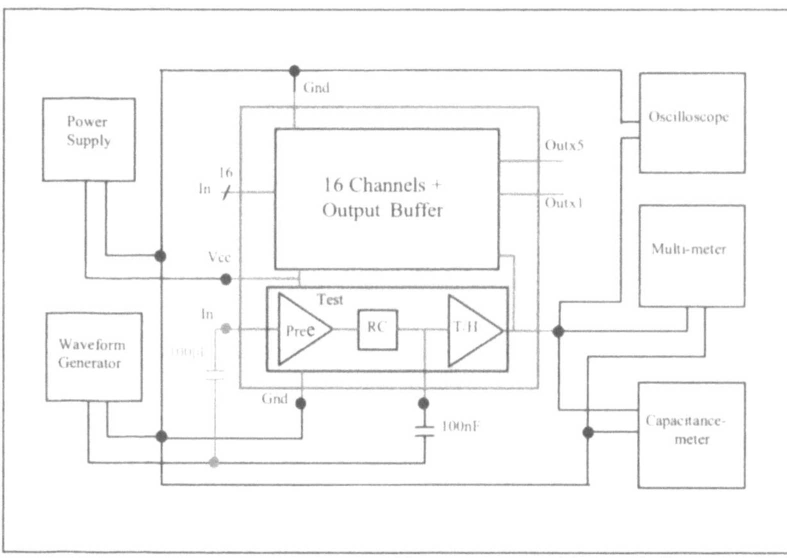

Figure 14. Measurements on the chip, in light gray the configuration used to determine the i/o response of the whole test channel

a) We used a capacitance-meter to estimate the value of the capacitance introduced by the probes; with the instruments from ST we found C=130pF, which is comparable with C_{par}; anyway this value is strongly dependent on the instruments (and not only, *i.e.*, the length of the cables, …).
b) Disconnecting the probe of the multi-meter during the measurement the large overshoot lowered.

The following pictures compare the experimental responses with the simulated ones using a 280pF capacitive load at the circuit output. Take the input step signal amplitude as a reference to evaluate the output signal overshoot.

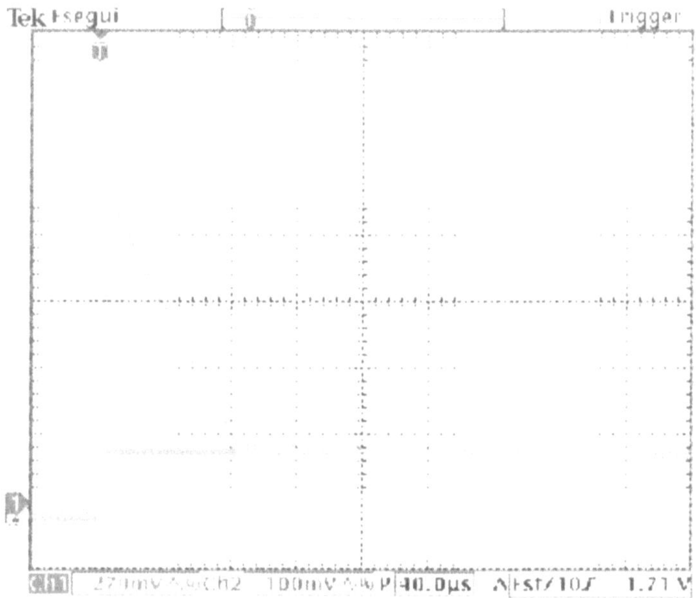

a

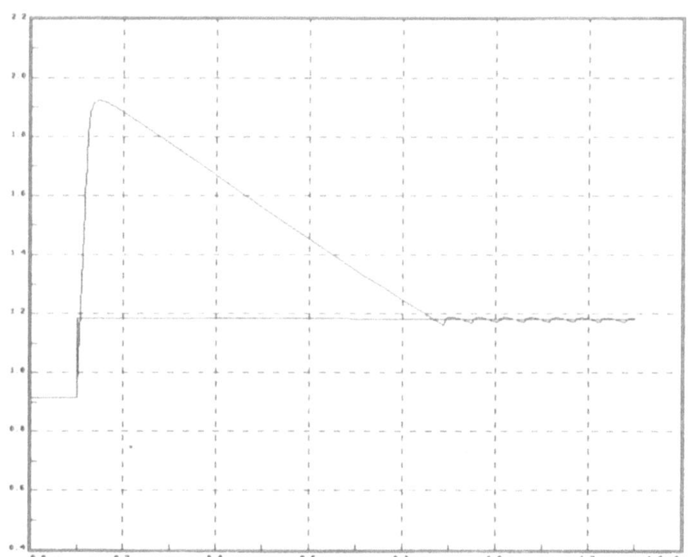

b

Figure 15. a. real and b. simulated track/hold i/o response. Vcc=5V, Vin=100mV pp.

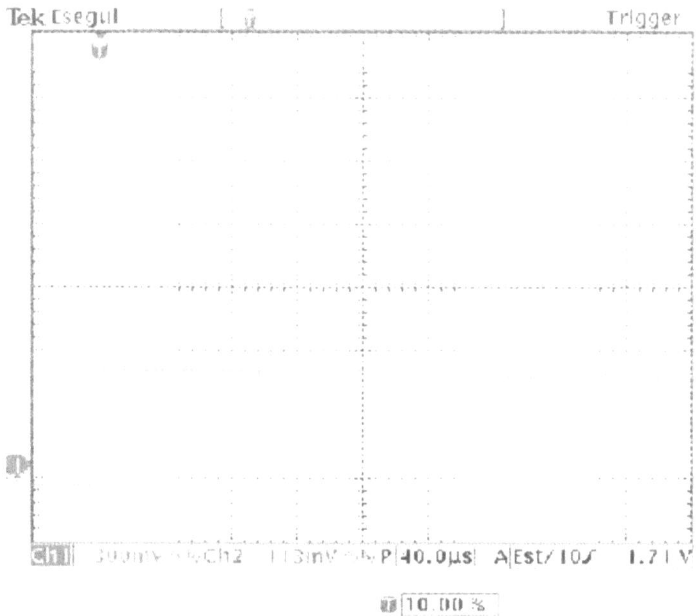

a

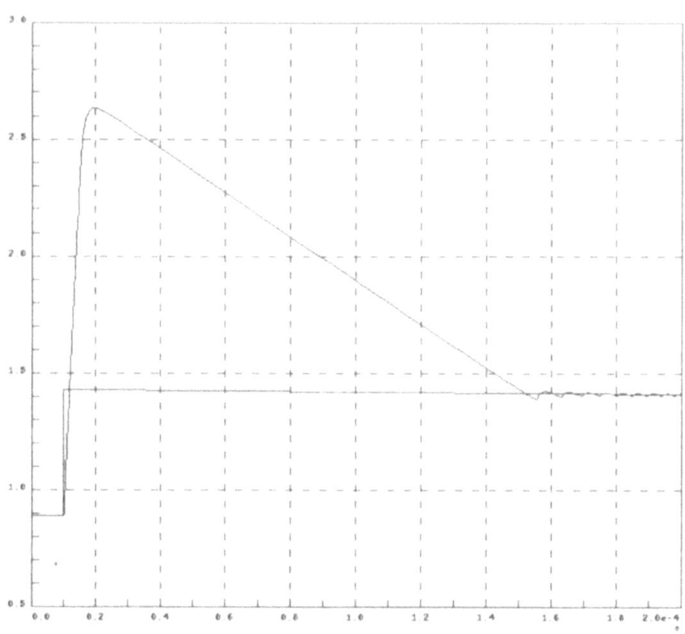

b

Figure 16. a. real and b. simulated track/hold i/o response. Vcc=3.5V, Vin=200mV pp.

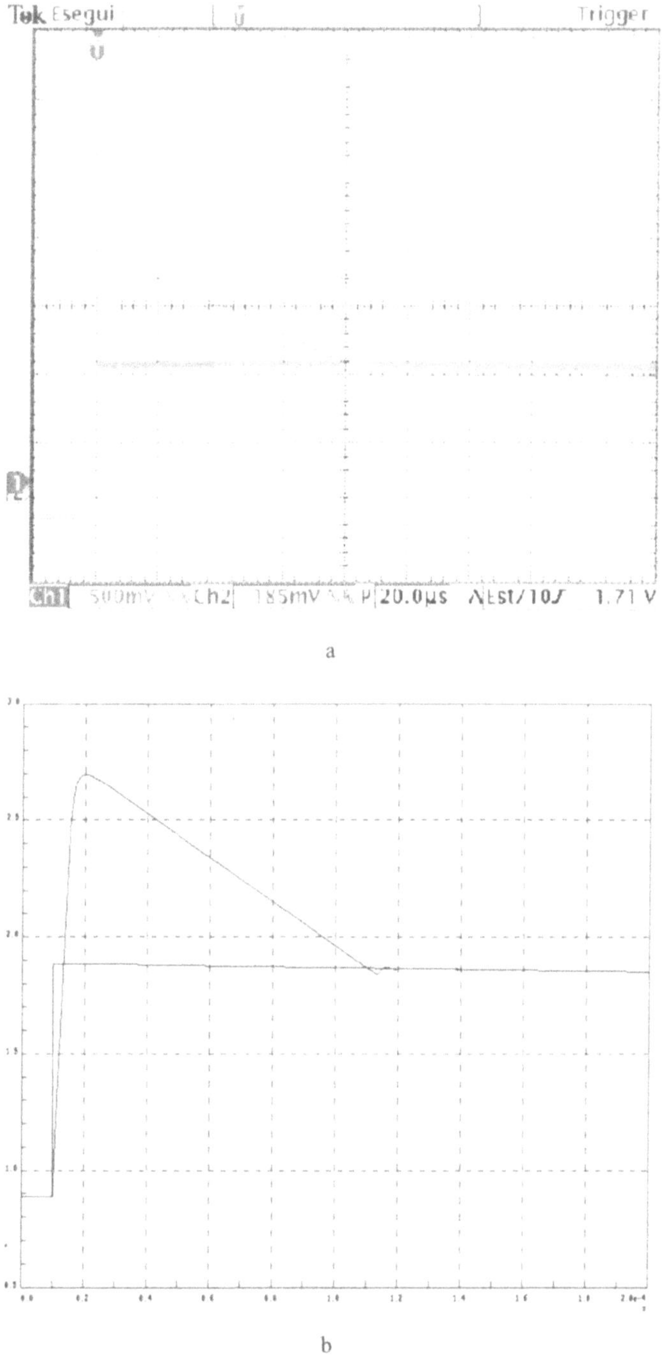

Figure 17. a. real and b. simulated track/hold i/o response. Vcc=3.5V, Vin=380mV pp

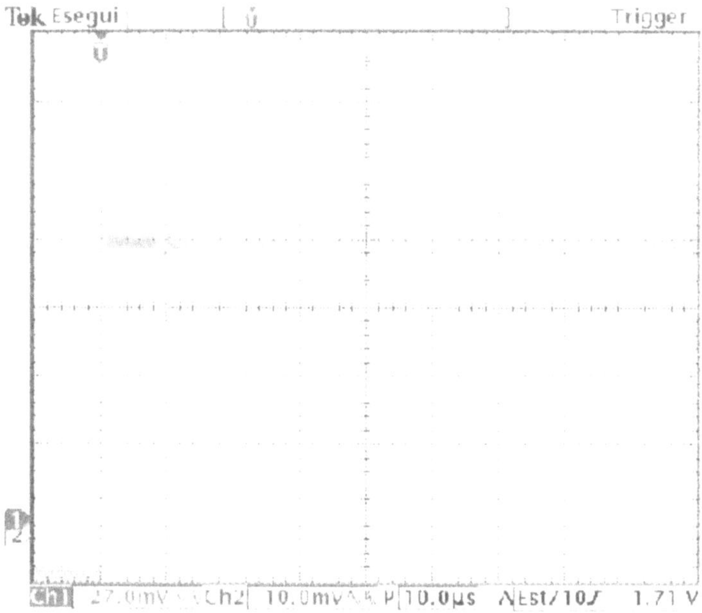

a

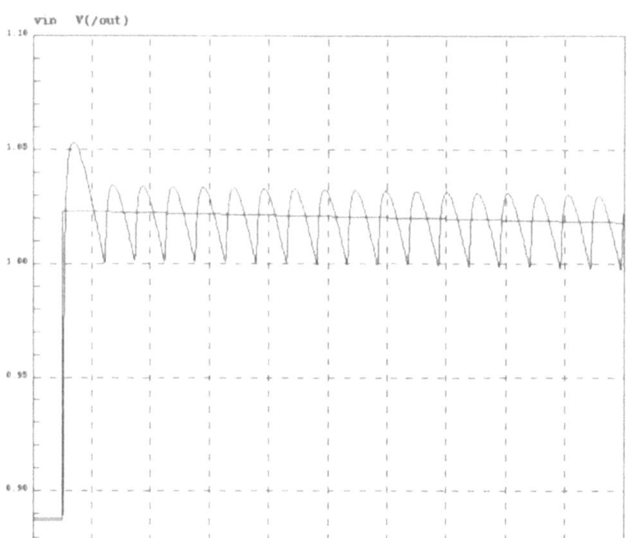

b

Figure 18. a. real and b. simulated track/hold i/o response. Vcc=3.5V, Vin=50mV pp

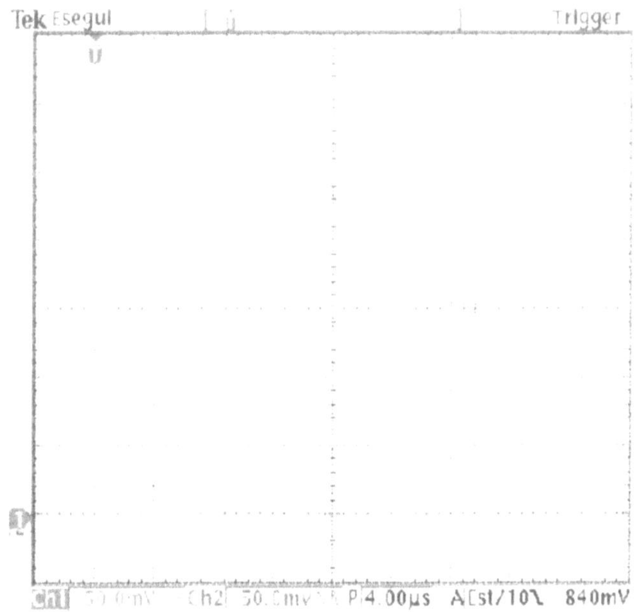

a

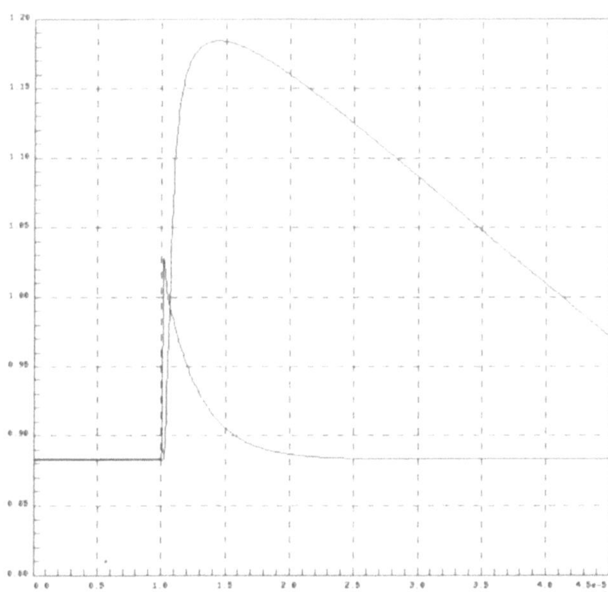

b

Figure 19. a. real and b. simulated test channel i/o response. Vcc=3.5V, Vin=190mV pp. The plot in cyan is the signal at the trans-resistance amplifier output.

Improved Version of the Track/hold Circuit

In this section we will present an improved version of the track/hold circuit that exhibits a somewhat higher open-loop gain and succeeds to meet the desired transient requirements on the hold-mode output signal. The idea is to use the shunt differential stage in the original circuit (see Figure 5), which contributed to the response only during a very brief time interval (from Trigger on till emitter follower Q_4 off) and in practice was not useful, to provide the circuit with a further gain stage (see Figure 20); frequency compensation is then performed by means of two capacitors C_{m1}, C_{m2} (nested Miller technique). Although this leads to a slightly reduced frequency bandwidth, it is not a serious drawback here since we are first interested in high accuracy.

With regard to the transient behaviour, we eliminated the compensating subcircuits that didn't work properly and replaced the bipolar followers Q_4, Q_5 with MOSfet ones (M_1, M_2). The low irradiation levels (<10Krad) the circuit is supposed to tolerate and the use of such devices only as followers make this solution suitable here.

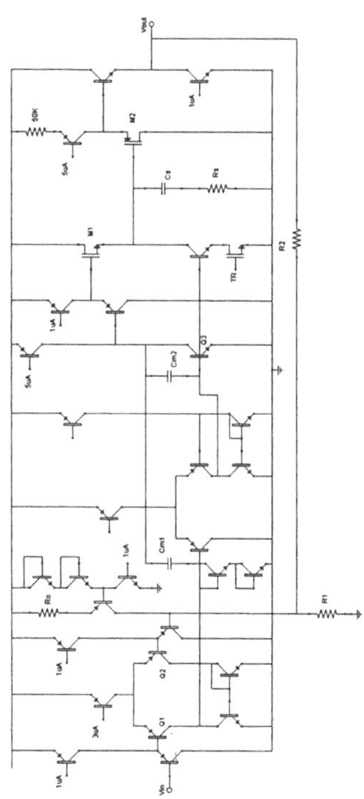

Figure 20. Improved track/hold circuit

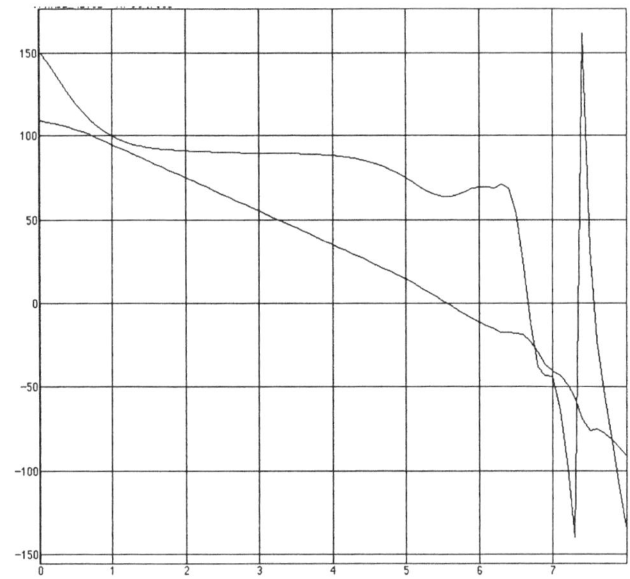

Figure 21. Frequency response of the modified track/hold circuit

Final Remarks

In conclusion, we would like to point out that the achievement of such results has not required the complete redesign of the circuit. This fact has a great impact on the fabrication cost and time reduction since the changes needed to improve the chip can be implemented by modifying only the last metal mask and not the whole set.

References

[1] P. Gray, R.G. Meyer. Analysis and Design of Analog Integrated Circuits.

[2] M. Dentan, CERN-EP-ATE-CEA. Radiation Effects on Electronic Components and Circuits.

[3] A 100-MHz 100db Operation Amplifier with Multipath Nested Miller Compensation Structure. IEEE Journal of S-SC, 27 (12), DEC 1992.

[4] Feedforward Compensation Techniques for High-Frequency CMOS Amplifiers. IEEE Journal of S-SC, 25 (6), DEC 1990.

[5] A CMOS Low-Distortion Fully Differential Power amplifier with Double Nested Miller Compensation. IEEE Journal of S-SC, 28 (7), JUL 1993.

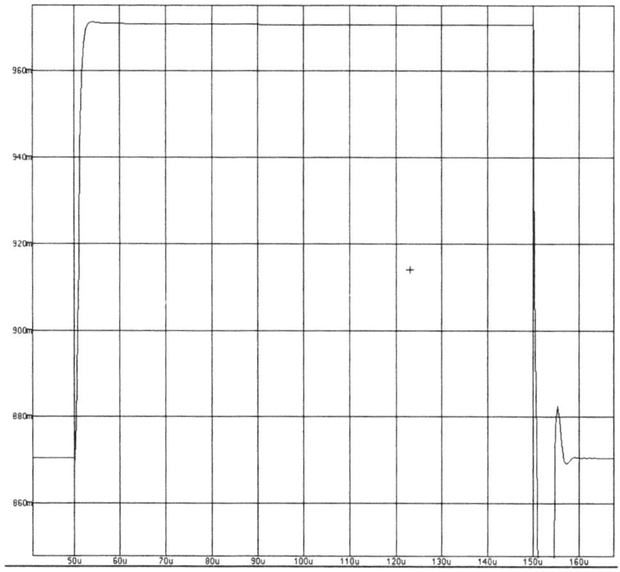

a

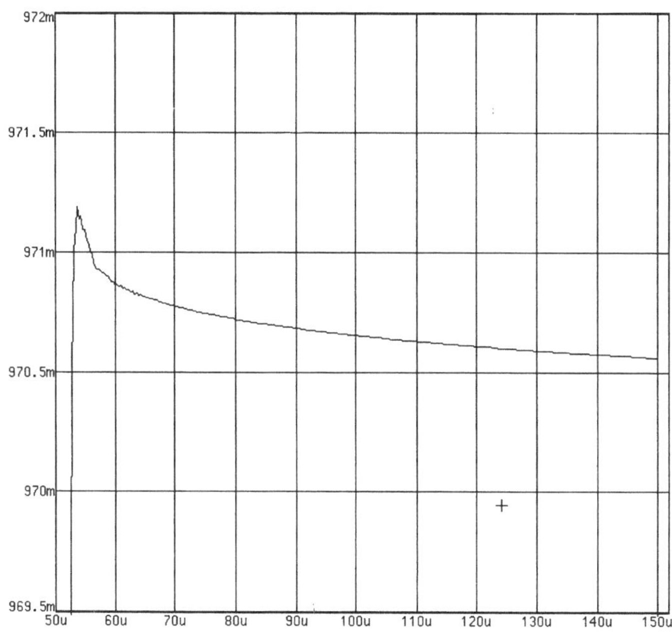

b

Figure 22. a. modified track/hold output response (hold mode); b. modified track/hold output response (hold mode). The overshoot and the voltage droop satisfy the desired requirements.

Circuit-level Implementation for a DC/DC Fuzzy Controller in BCD3s Technology

Giuseppe Vanacore

STMicroelectronics S.r.l.
Catania site – Italy

e-mail: giuseppe.vanacore@st.com

Introduction

DC/DC converters are broadly used in mobile phones, hard disks, computers, and in many other portable devices, where the increasing demand requires products to meet higher targets.

The intense research activity on integrated converters, is basically bound to improve transient performances such as a faster response to load variations along with wide range stability.

The idea of improving performances lies behind all the fuzzy-based control systems, algorithms whose integration (in BCD3s technology, as for here) delivers all the well-acknowledged perks related to this new control methodology (cf. for instance [3] and [6]).

Moreover, when it comes to circuital implementation, both analog and digital, what mostly must be taken into account are supply voltage fluctuations and parametric variations due to temperature shifting.

The present activity, developed thanks to the support of the Switching-Regulators Power-Conversion Group at STMicroelectronics in Catania, deals with the electrical design and characterization of the main blocks for what is known in literature as a *step-up* (*boost*) *converter* at fixed frequency with (fuzzy) duty cycle control (PWM controller) [7].

Key Words and Principles of Operation

Voltage regulators make an input power (usually from a not regulated voltage generator) available to the output (within a given range of supply variation) regardless (to some extent) of any load variations. The commonest families are: series linear regulators and switching regulators. In most cases, duty cycle control is achieved out of a simple current-mode scheme (linear regulation).

Switching regulators cover a wider control spectrum, up to higher performances (80%—90%) and allowing to lift up the input voltage (*step-up configuration*).

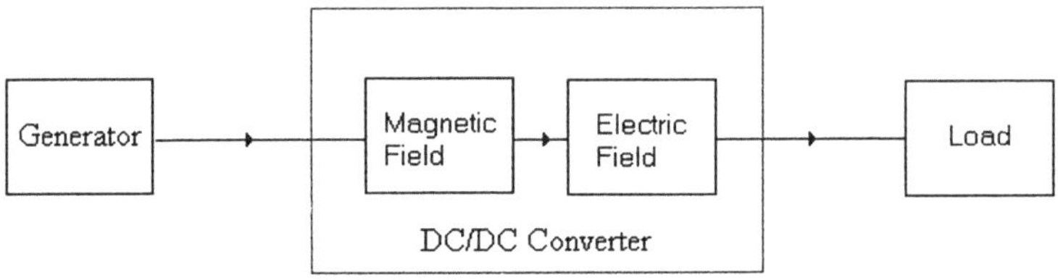

Figure 1. Energy Flow: a converter-regulated *power-load*

DC/DC converters are a key point in energy transfers between the mains and the load, firstly storing energy in the magnetic field of an inductor then transferring it to the electric field of a load-parallel capacitor as in Figure 1.

Figure 2 outlines the *step-up* topologies. Two passive elements come up with a bipolar transistor (SW1) and a diode (SW2), which account for power switches. Inductor charging and subsequent energy transfer to the capacitor are timed by a suitable signal (clock), designed to keep output voltage constant with respect to the absorbed load current.

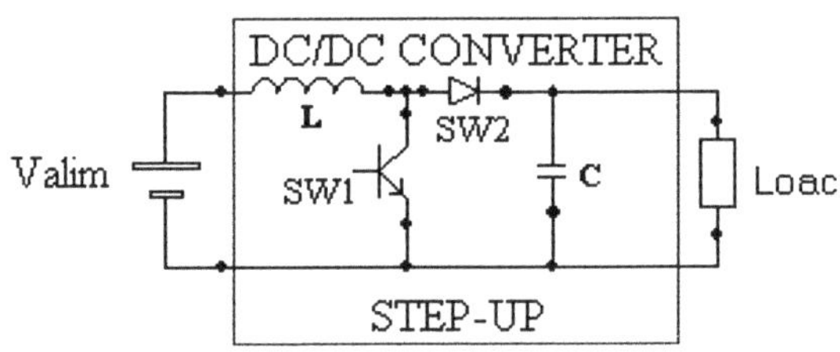

Figure 2. Circuit scheme of the step-up topology

In every T length commutation cycle, three intervals can be identified, depending on the state of the switches: T_{ON} (transistor-on time): transistor *SW1* is ON and diode *SW2* is OFF. During this time energy is stored in the inductor magnetic field; T_D (diode-on time): transistor *SW1* is interdicted by the control signal while diode *SW2* is ON; in this phase energy stored in the inductor decreases, due to energy transferring from the magnetic field to the capacitance electric field; T_I (idle time or freewheeling time): *SW1* and *SW2* are both OFF, all energy required by the load is supplied by the capacitance, whose voltage decreases, while both inductor current and its magnetic field are zeroed. The three time intervals add up to the overall switching period: $T = T_{ON} + T_D + T_I$.

Presence or not of T_I determines two possible operation modes:

continuous mode (CCM), where inductor current, and therefore its magnetic field, never zeroes in the whole cycle ($T_I = 0$);

discontinuous mode (DCM), in which inductor current comes to zero ($T_I \neq 0$) during the switching cycle.

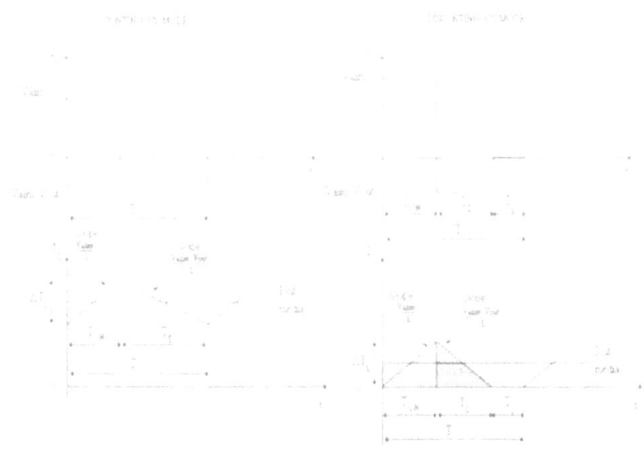

a b

Figure 3. Step-up: current and voltage wave forms in a. CCM; b. DCM

Comparison between CCM and DCM Mode in Step-up Converters

Step-up converters in CCM mode are characterized by a control scheme leading, under particular load conditions, to instability, large output ripple and sub-harmonic oscillations [3].

On the other side, a DCM step-up converter offers good stability performances but, depending on the load current, may give rise to larger peak currents through switches and reactive components.

The better DCM stability response makes it more popular and advantageous than the CCM but the consumption and the sizing of the components in battery-powered systems jostle more and more for highly sophisticated step-up converter controllers fitted to work without instability concerns in both CCM and DCM mode.

The present task aims to improve the layout definition and integrated realization of the above-mentioned controller.

The overall schematic plans are shown in Figure 4a and 3b. Here, 9 main blocks can be identified:

a) oscillator;
b) reference voltage block;
c) ramp (saw tooth) generator;
d) sense amplifier block;
e) comparator;

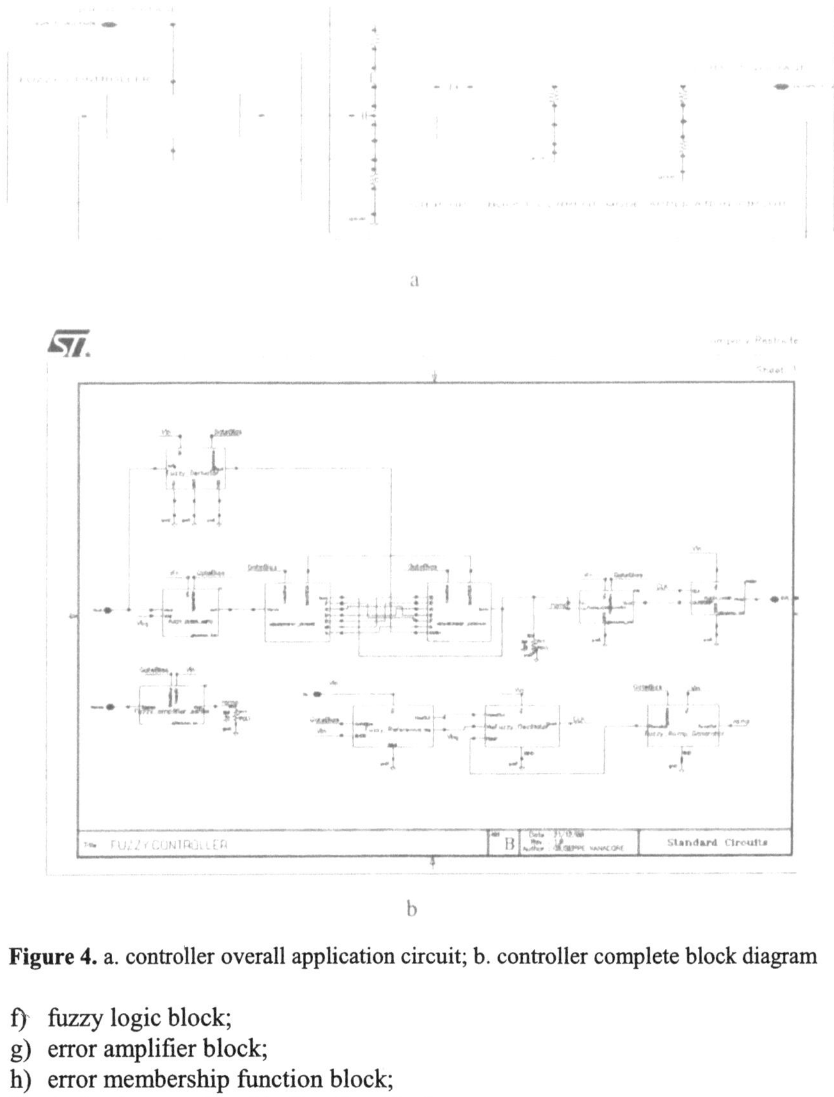

Figure 4. a. controller overall application circuit; b. controller complete block diagram

f) fuzzy logic block;
g) error amplifier block;
h) error membership function block;
i) derivative membership function block.

Prior to giving further details of each block, it is useful to give hereafter a general overview around the meaning of the control strategy at issue.

Control Techniques for DC/DC Converters

PWM Control: Voltage-mode and Current-mode

A DC/DC converter is intended to supply a load at a constant voltage in spite of power voltage and load current variations. To meet this target, it is necessary a closed loop control system, where a logical signal would take over the activation of power switch SW1 (see Figure 2) dosing the energy flowing to the load. This control, if achieved modulating the duration T_{ON} of the power switch, being T fixed for a given switching frequency $f_s = \frac{1}{T}$, is said PWM, that is, duty-cycle control $d_1 = \frac{T_{ON}}{T} = T_{ON} \cdot f_s$.

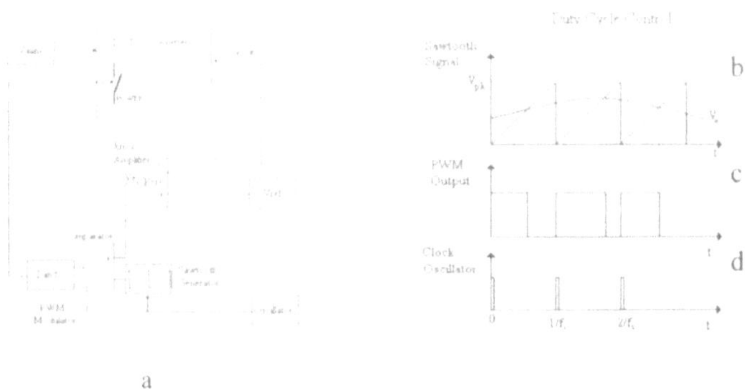

Figure 5. a. block scheme of a PWM controller; b. main shape waves

In Figure 5a the simplified block scheme for a PWM controller in his general form is shown. It is possible to identify: an oscillator that determines the time frequency f_s of the power switch and an error amplifier that produces a signal V_e as:

$$V_e = K \cdot (V_{REF} - V_{OUT}) \tag{1}$$

where K is the static gain of the amplifier, V_{OUT} the regulated output voltage and V_{REF} the reference voltage inside the regulator.

The block scheme consists, besides, of a comparator that correlates the error signal V_e and a periodic signal of period $\frac{1}{f_s}$ (see Figure 5b). The output of the comparator is driven, along with the clock signal (Figure 5d), to a latch whose output (see Figure 5c) determines the switching on and off of the power switch.

If the constant signal comes up always with the same $\frac{dV}{dt} = const$, for example using the clock-generating oscillator, the PWM control is said to be *voltage-mode*.

If rather, the signal has a finite slope $\frac{dV}{dt} = \alpha \cdot \frac{dI_L}{dt}$, α being a constant of proportionality and I_L the inductor current, the PWM control is said to be *current-mode* [2].

a b

Figure 6. Block scheme of a PWM current-mode controller and main shape wave

Figure 6a shows the block scheme for a PWM current-mode control regulator in which two feedback rings are recognizable. In the outer ring, the control of output voltage takes place, in the inner one the information about inductor current [1] is used [4]. This type of regulators uses a saw tooth signal V_{sense} proportional to the inductor current I_L:

$$V_{sense} = \beta \cdot I_L \cdot R_{sense} \tag{2}$$

where β is a constant of proportionality and R_{sense} a small value resistance in series to the switch, run through by the inductor current I_L during the interval T_{ON}.

A control technique frequently used in current-mode is the peak current regulation [2]. Briefly, power switch is turned off when the voltage signal proportional to inductor peak current V_{sense} gets to the value of the output signal V_e of the error amplifier as in Figure 6b: in this way, it is possible to have a faster response to both input voltage and output current variations, gaining an intrinsic limitation for peak currents against output overloading.

One of the drawbacks of current-mode control is the intrinsic instability of the closed loop system for values $d_1 > 0.5$ [4]; this may result in sub-harmonic oscillation of V_{OUT} and perturbations of the current. This limitation can be overcome subtracting a *compensation* voltage ramp with constant slope m from the error voltage V_e.This technique is known as *slope compensation* [2], [5].

In a bare current-mode control, a perturbation $\Delta_0 I_L$ of the inductor current sets off a perturbation $\Delta_1 I_L$ greater than $\Delta_0 I_L$, in the next clock period, if $d_1 > 0.5$ (Figure 7b) or smaller if $d_1 < 0.5$ (Figure 7a), the link betwixt them in two consecutive clock periods being:

$$\Delta_1 I_L = -\Delta_0 I_L \cdot \frac{m_2}{m_1}. \tag{3}$$

Without *ramp of compensation*, condition $|m_1| > |m_2|$ is equivalent to $d_1 < 0.5$, the value 0.5 ensuring intrinsic stability.

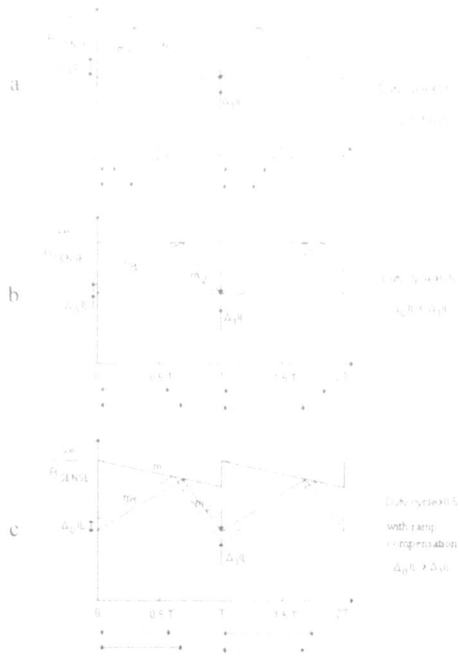

Figure 7. Instability for a PWM current-mode converter

In the case $d_1 > 0.5$ system stability can be ensured subtracting from V_e a voltage ramp signal of angular coefficient m, as in Figure 7c, in such a way that condition (3) holds:

$$\Delta_1 I_L = \Delta_0 I_L \cdot \left(\frac{-m_2 + m}{m_1 + m} \right) \tag{4}$$

or

$$m > \frac{m_2 - m_1}{2} \tag{5}$$

For a step-up regulator, the most critical situation (*i.e.*, when $d_1 = d_{1_{max}}$) in (5) becomes [1]:

$$m_{\min} = \frac{V_{ALIM} \cdot \left(2d_{1_{max}} - 1\right)}{2L \cdot \left(1 - d_{1_{max}}\right)} \tag{6}$$

Therefore, by choosing a compensation ramp with $m \leq m_{\min}$, in every cycle the previous perturbation alleged to bring on system instability is dramatically reduced.

A disadvantage of current-mode control technique is the extreme noise sensibility: $R_{sense} \cdot I_L$ voltage, used to generate the saw tooth signal, V_{sense}, is very small (3 or more magnitude orders) compared to the level of voltage V_e, so that a small noise could modify the turning off instant of the power switch (jitter-noise).

Such problems strongly limit practical applications of step-up converters in continuous mode operation. Thus, the restrained possibility to control this type of systems with linear control techniques leads to look for better solutions, less susceptible to system contour conditions and their variations, within non-linear control techniques. Among the possible approaches of non linear control, particular interest has arose around fuzzy logic, partly explained by a straightforward controller synthesis, based on pouring phenomenon knowledge out to a set of linguistic rules describing the expected behaviour in different operative conditions.

Synthesis of the Fuzzy-based Control System

The idea is to integrate a standard *ramp compensation* current-mode control within a fuzzy logic algorithm that acts on system transfer function to make up for the negative effects of load variations via a non linear regulation. The aim is to join the advantages of current-mode control, with the flexibility of a fuzzy control system in the voltage loop, widening the range of stability in comparison to the traditional control. The reference schematic adopted is in Figure 8.

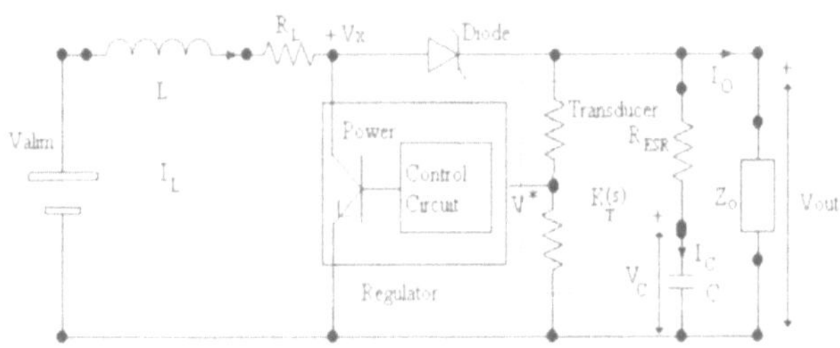

Figure 8. Reference schematic

The error and its first derivative (D_E) are inputs to the fuzzy algorithm.

Figure 9. Schematic of the adopted control

The *error amplifier block* of the linear regulator has been replaced with a fuzzy control block whose inputs are the error and the first derivative of the output voltage. The output of this block is one of the two inputs of the comparator (the actual PWM modulator). The goal is to act on converter transfer function in a non linear mode by compensating the negative effects of the linear control according to the principle scheme visible in Figure 9.
As for the fuzzy algorithm, linguistic rules are empirically derived from the following criterions:

- for a small error in absolute value, the output of the fuzzy block ('*F_Out*') is depending on load condition ('*Derived*');
- an elevated and positive error value forces *F_Out* to its maximum value;
- an elevated and negative value of the error forces *F_Out* to its smallest value.

For more details about the fuzzy controller, compare [7].

Circuit Blocks Description

As stated previously, the whole circuitry has been developed in BCD3s technology. This BiCMos technology is characterized by a 1μ mosfet channel length, 5/13V mosfets and 30V bipolar transistors.

Oscillator

This is the clock generating block. The circuit is a typical relaxation oscillator. The output wave form is resulting from charge/discharge of two capacitors (say C1 and C2), the first one 4pF's and the other 19.5pF's.

The biasing current being drawn by a 18uA current mirror, the charge of the capacitance is simply given by:

$$I = C_i \Delta V / \Delta T_I \qquad \text{for } i=1,2.$$

As ΔV is 1.24 V (from reference block) for both, the overall period is:

$$T = \Delta T_1 + \Delta T_2 = (C_1 + C_2)\, \Delta V / I$$

so that the highest switching frequency is inversely proportional to the sum $(C_1 + C_2)$ while the ratio $C_1/(C_1 + C_2)$ gives the peak duty cycle:

$$\delta_{\max} = C_1 / (C_1 + C_2)$$

The reason for considering $\delta_{\max}$ lies in the liaison between δ and the desired V_{in}/V_{out} ratio for a step-up converter:

$$\delta = 1 - V_{in}/V_{out}$$

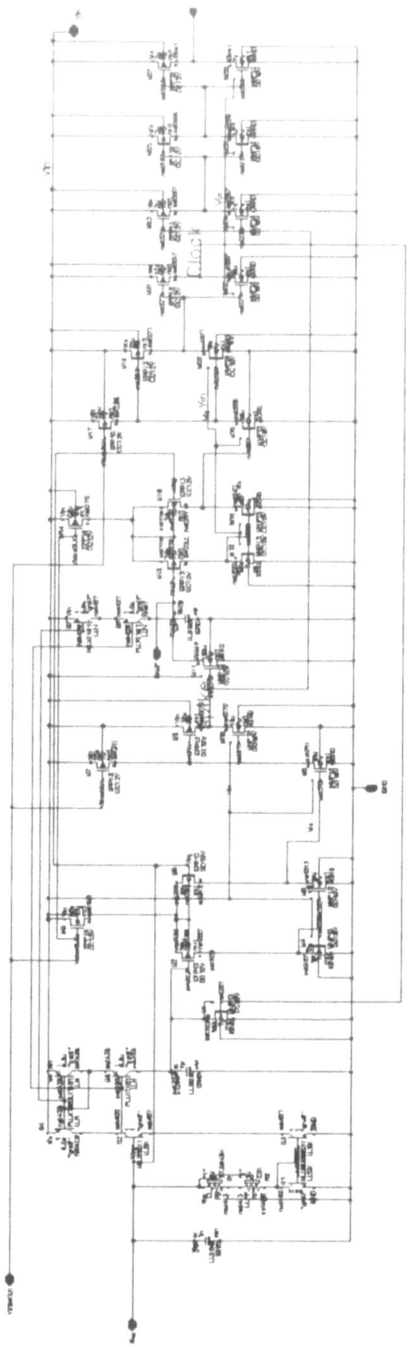

Figure 10. Oscillator schematic view

so that δ_{max} relates to the lowest V_{in} .

The circuit has been thermally compensated, as the VKL for the charging current holds:

$$I = (1.24 - V_{be})/(R_1 + R_2)$$

V_{be} being the base-emitter voltage of the diode-connected npn transistor, that decreases with about –2.2mV/°C.

The compensation is obtained introducing a series of two resistors that behave differently with changes in temperature, the bigger one with a smaller temperature coefficient (about 900 ppm/°C^{-1} for a P-body resistance) and the smaller one with about 1250 ppm/°C^{-1}).

The oscillator output, with the aforesaid values, consists of a 160kHz rectangular wave with a $\delta_{max} = 0.85$; in Figure 11, there is also a spike signal (on the clock falling edge) required to properly trigger up the PWM logic latch.

Reference Voltage Block

This block yields a 1.24 V reference voltage (see Figure 12) necessary to determine error voltage in the error amplifier.

Band-gap circuits [8] utilize the decrease of V_{be} with temperature and the opposite behaviour of $V_T = K\ T/q$.

The band-gap generating circuit is preceded by a 18uA biasing circuit, in which the resistance is given by:

$$R_b = 2V_T \ln 3 \ / \ I_{bias}$$

where $I_{bias} = 18\text{uA}$, $V_T = 26\text{mV}$ and choosing three times one or another npn emitter area.

Once again, this resistance value is split into two resistors of different composition to achieve temperature compensation of I_{bias}.

The band-gap voltage also needs compensating as we have:

$$V_{band\text{-}gap} = V_{be_6} + 2R_4 I_{Q6} = V_{be_6} + 2R_4 V_T \ln 10/R_3$$

compensation calls for:

$$\partial V_{band\text{-}gap}/\partial T = 0 \ \ @ \ T = 25°C = 298K$$

thus constraining the ratio $R_4 / R_3 \simeq 5$.

Capacitance C_1 introduces a compensation dominant pole in the loop transfer function by Miller effect.

The reference block also has a shut-down pin to disconnect the load from the regulator hence minimizing power in stand-by conditions.

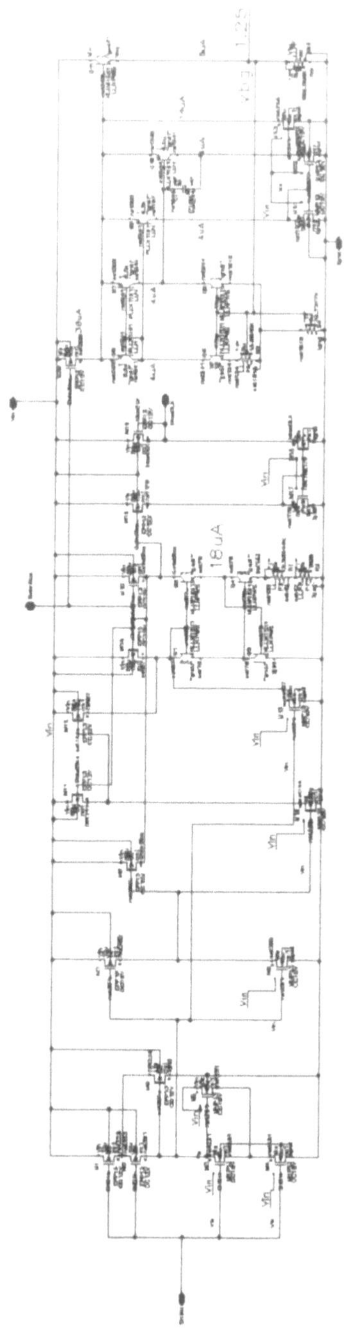

Figure 11. Reference voltage schematic block

Saw Tooth Generator

A saw tooth signal is required to apply ramp compensation when comparing load current to error voltage, as previously detailed. Though a signal like this could have also been derived from the oscillator circuit, this could have led up to dynamic misbehaving due to additional capacitive load. Hence this block is basically a level shifter (whose input is a voltage signal equal to the C1 ramp signal) adding a voltage controlled current source to the above-mentioned spike-like signal and whose output is a ramp current.

For this block, oversized Mosfets have been used in order to lower the difference V_{GS} - V_{th}, being the output voltage $V_{in} + V_{GS_p} - V_{GS_n}$.

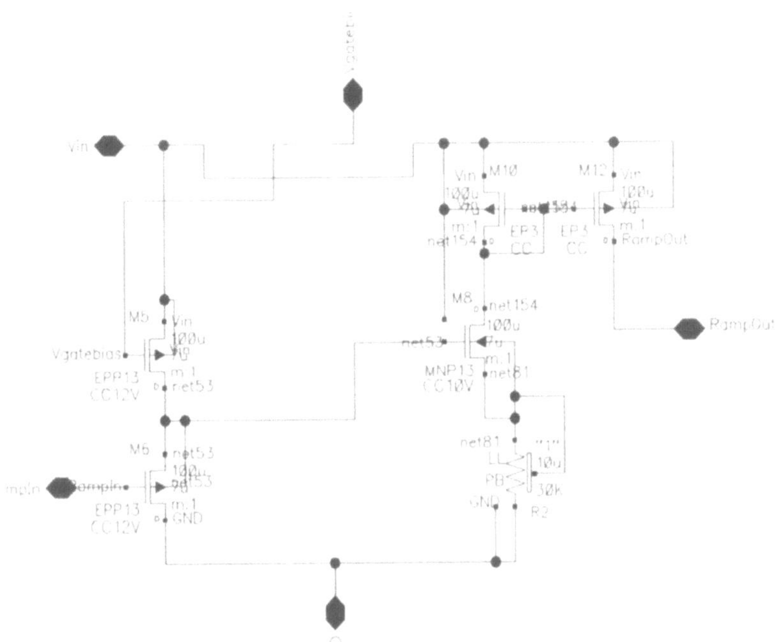

Figure 12. Ramp generator

The output current is simply given by:

$$\frac{V_{ram}(C1)+V_{GSp}-V_{GSn}}{30\text{k}\Omega}$$

Sense Amplifier Block

This block detects and amplifies the load current to the level required to be compared to the error voltage in the comparator block. The output is a current that adds to the saw tooth current (Figure 13).

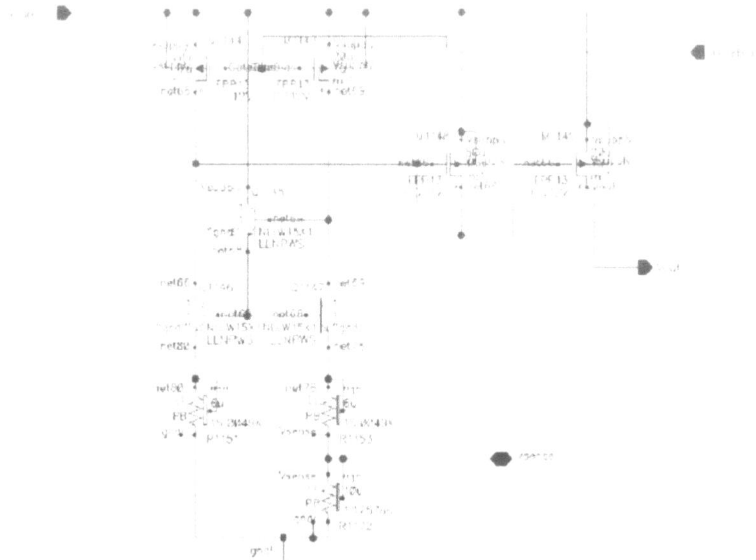

Figure 13. Current sense amplifier

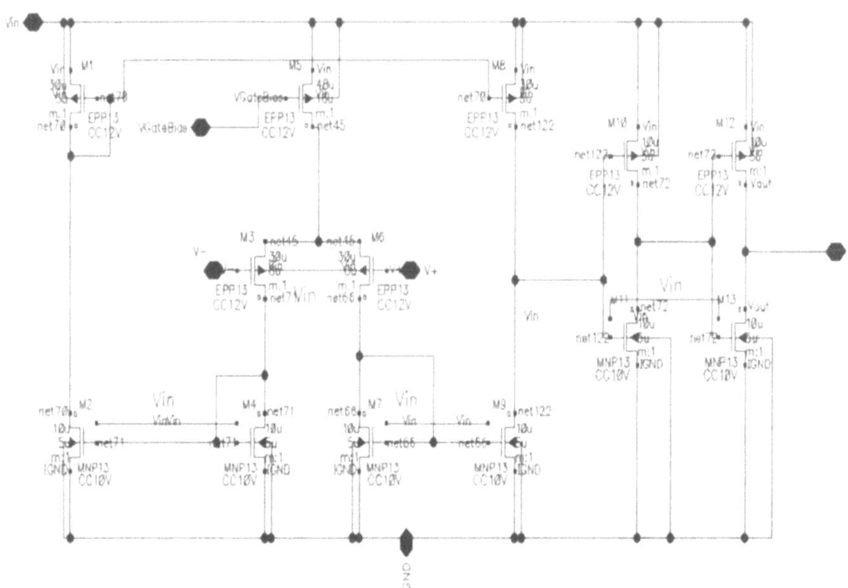

Figure 14. Comparator

The output current is given by:

$$I_{out} = V_{sense}/15\text{k}\Omega * 20/50$$

Comparator

This block compares the error voltage coming from the fuzzy logic conditioning block to the saw tooth voltage proportional to the switching current plus the compensation ramp, featuring a wide band PWM generator block (Figure 14).

Error Amplifier Block

Here output voltage is compared to reference voltage before being processed by the fuzzy logic analog block.

It is essential to have a good dynamic behaviour to avoid instability (high static gain with a sharp decrease at higher frequencies, so that the phase margin zeroes when the amplitude margin is sufficiently low).

This implies a different sizing of components from the comparator's and a wide band topology (Figure 15).

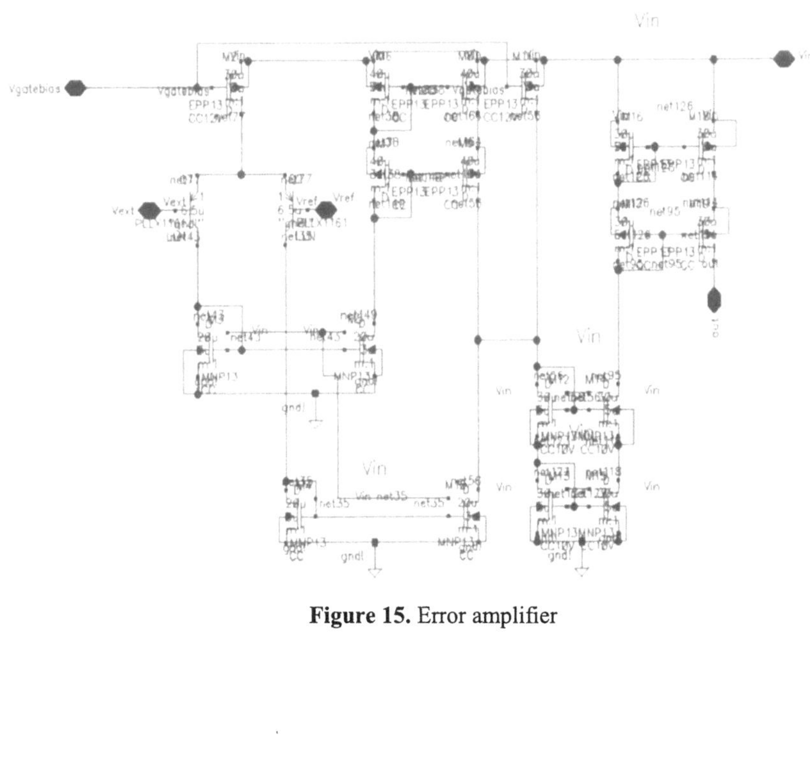

Figure 15. Error amplifier

Membership Function Blocks and Fuzzy Logic Block

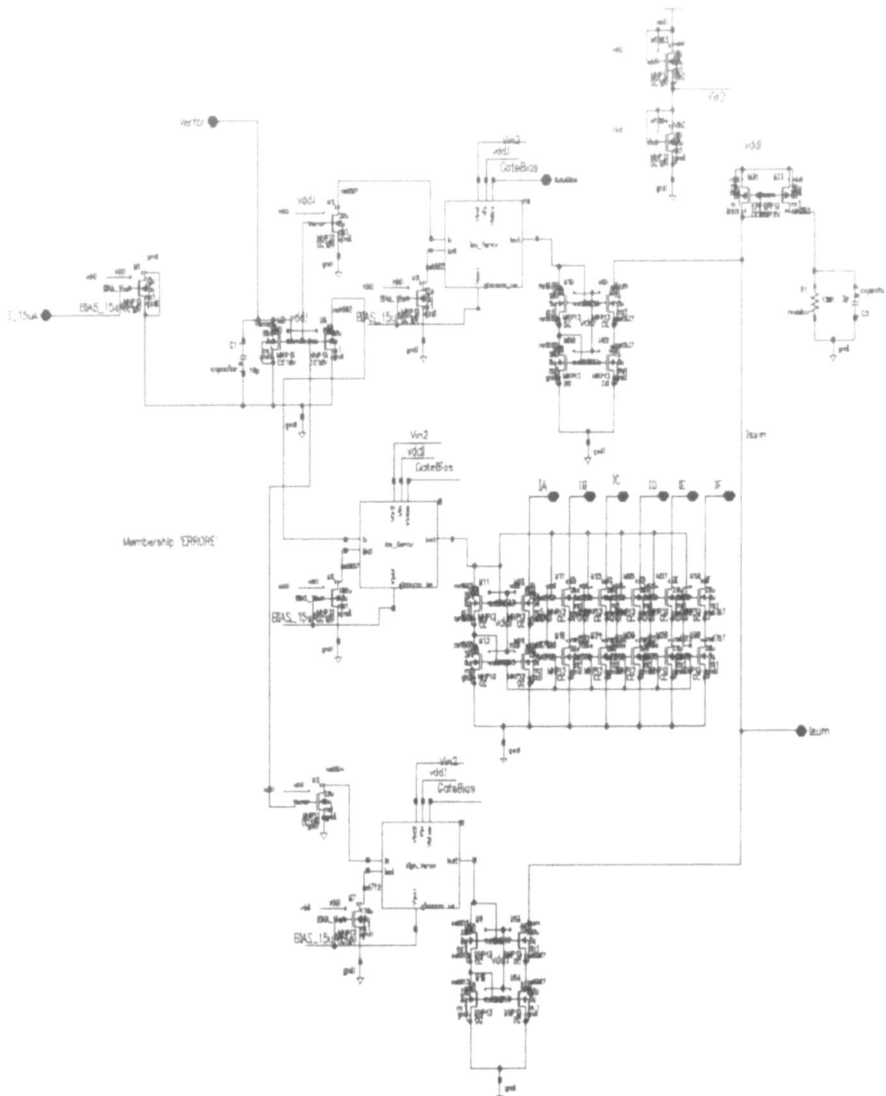

Figure 16. Error membership block

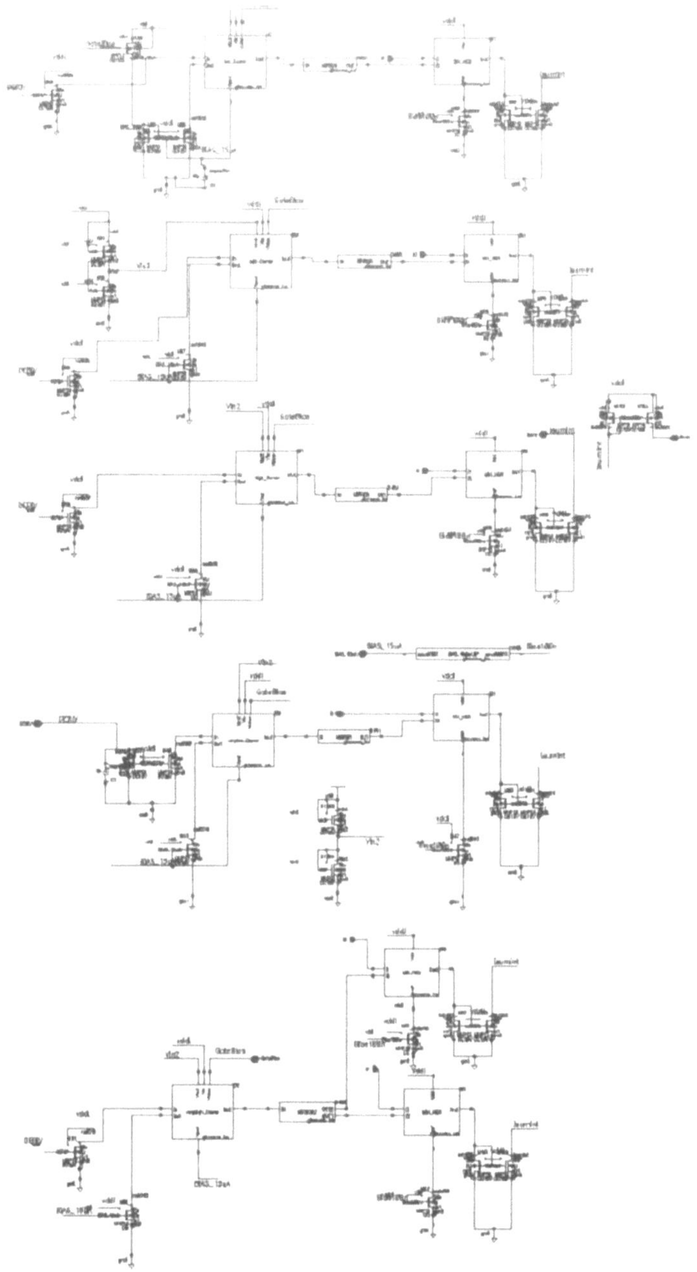

Figure 17. Derivative membership block

These blocks implement the logic algorithm, previously discussed, the output of which switches the power on and off, in circuits. The fuzzy algorithm sets the value to be compared with the saw tooth proportional to inductor current in the PWM comparator. A sufficient positive value of V_E ($V_{OUT} << V_{NOM}$) delivers the maximum output from the fuzzy block; similarly, an amply negative value of V_E (V_{OUT} >> V_{NOM}) holds the least output.

The input linguistic variables are:

Error (E): expression of the output voltage error (V_E) to nominal voltage (V_{NOM}), with a range of three linguistic values (*Error neg, Error zero, Error pos*).

Derivative (DE): expression of "error derivative" (DE) of the output voltage, within a range of five linguistic values (*Very Small, Small, Medium, Big Very Big*).

Every linguistic rule needs a comparator block to determine its activation degree. The algorithm also needs a minimum block (Figure 18) which picks the smaller output current.

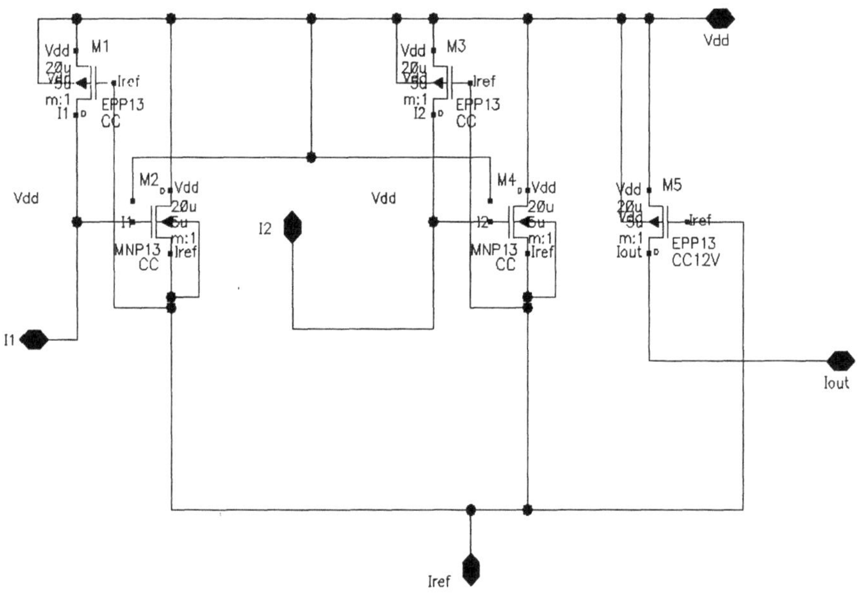

Figure 18. Minimum block

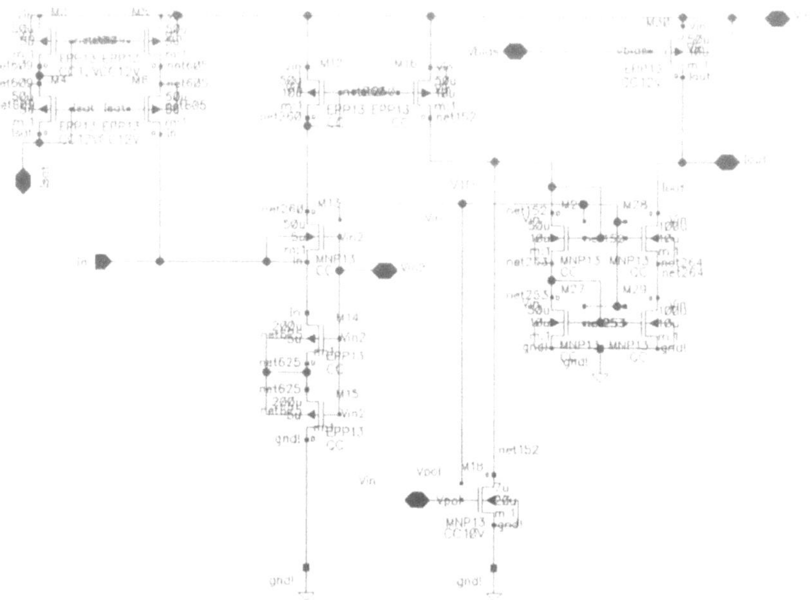

Figure 19. Sample membership function block

Figure 19 shows a sample block used to implement the activation degree detecting of a membership function: currents I_n and I_{sat} have different ways and their difference is transferred to the output.

The defuzzified output is the input to the comparator block and is the weighed sum of the activation degrees for every rule.

The rules are implemented via current mirrors with different mirror factors. The last block is the driver circuit of the power switch.

Conclusions

The preliminary layout of the regulator is shown hereafter. The die sizes 1.8*1.8 mm in a SO8 package. The aim of the activity was to develop circuital blocks for a specific BiCMos controller featuring fast load detecting and robustness to supply variations by virtue of a fuzzy algorithm. The block approach, commonly used for digital applications design, has proved useful here as well, mainly for its simplicity and the opportunity to reuse many times a reduced set of components in an analog environment.

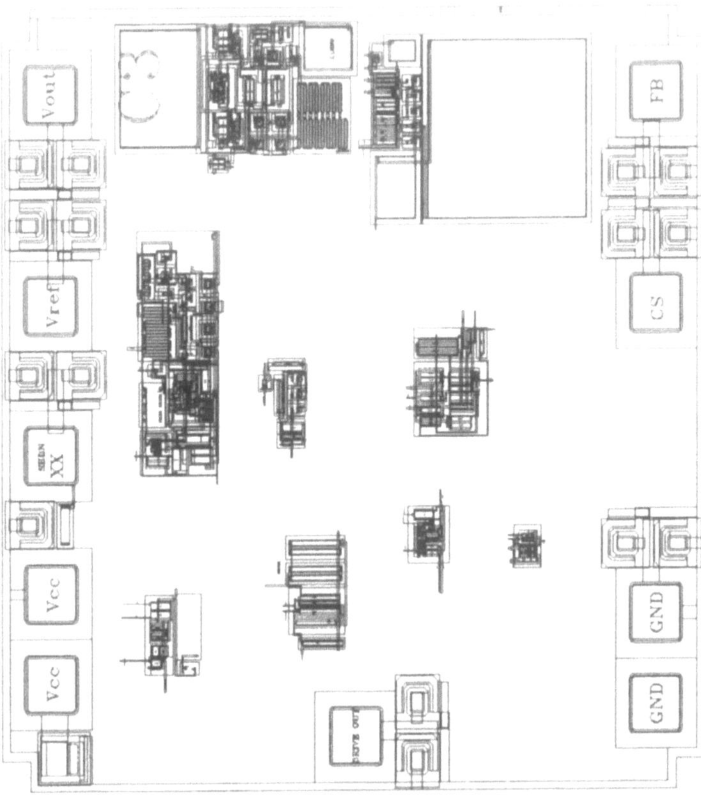

Figure 20. Preliminary layout

References

[1] D. M. Mitchell. DC/DC switching regulator analysis. McGraw-Hill.

[2] B. Murari, F. Bertotti, G. A. Vignola. Smart power ICs-technologies and applications. Springer 1996.

[3] A. Lionetto. Controllo current-mode in logica fuzzy per regolatori a commutazione in configurazione step-up. Tesi di Laurea, Dipartimento Elettrico, Elettronico e Sistemistico, Università degli Studi di Catania.

[4] C. Burgio, P. Cammelli, P. G. Maranesi, V. Varoli. The DC loop-gain of PWM voltage regulators. IEEE PESC Rec., 1986.

[5] Linear Data Book. Modelling, analysis and compensation of the current-mode converter. Unitrode Application note, 1995.

[6] G. Rizzotto, M. Lavorgna, M. Lo Presti. Metodologie per la sintesi e l'analisi di controllori fuzzy. Cavallotto Edizioni 1996

[7] G. Scalia. Design of a switching regulator with fuzzy control in BCD technology. Scuola Superiore di Catania, Advanced Course a.y. 1999/2000.

[8] P.R. Gray, R.G. Meyer. Analysis and design of analog integrated circuits. McGraw-Hill

Realization of a Place and Route Flow Using PKS, a New Cadence Design System Tool

Giancarlo Savarese
Salvatore Geraci

Accent Design Technology Services S.r.l.
Via F.Gorgone 6, 95030 Catania - Italy

e-mail: rsavarese@tiscalinet.it
gssalva@inwind.it

Introduction

The aim of the stage was to realize a complete Place and Route flow of a block of the integrated circuit shown in the next page, using PKS (Physical Knowledgeable Synthesis), and compare the results with those obtained with SE (Silicon Ensamble).

The block chosen is the MCM, because it has a shape of particular interest due to the presence of corners, which are critical points in terms of routing.

Standard flow is performed by Silicon Ensamble tool, while new flow is performed by PKS tool.

Standard flow needs only SE to be performed, while new flow needs both SE and PKS.

The reason is that the version of PKS used is not the final one, and so it is necessary to exit from PKS and enter SE to generate the clock tree and to execute the final routing.

The characteristics of MCM module are:

- dimension: 3300 × 4920 μm;
- Frequency: 25 MHz
- Components: 79,000
- Instances: 200,000

The major advantage of using PKS is that the tool achieves, during the first placement, an optimization of the logic functions of the block. This results in a shorter optimization step.

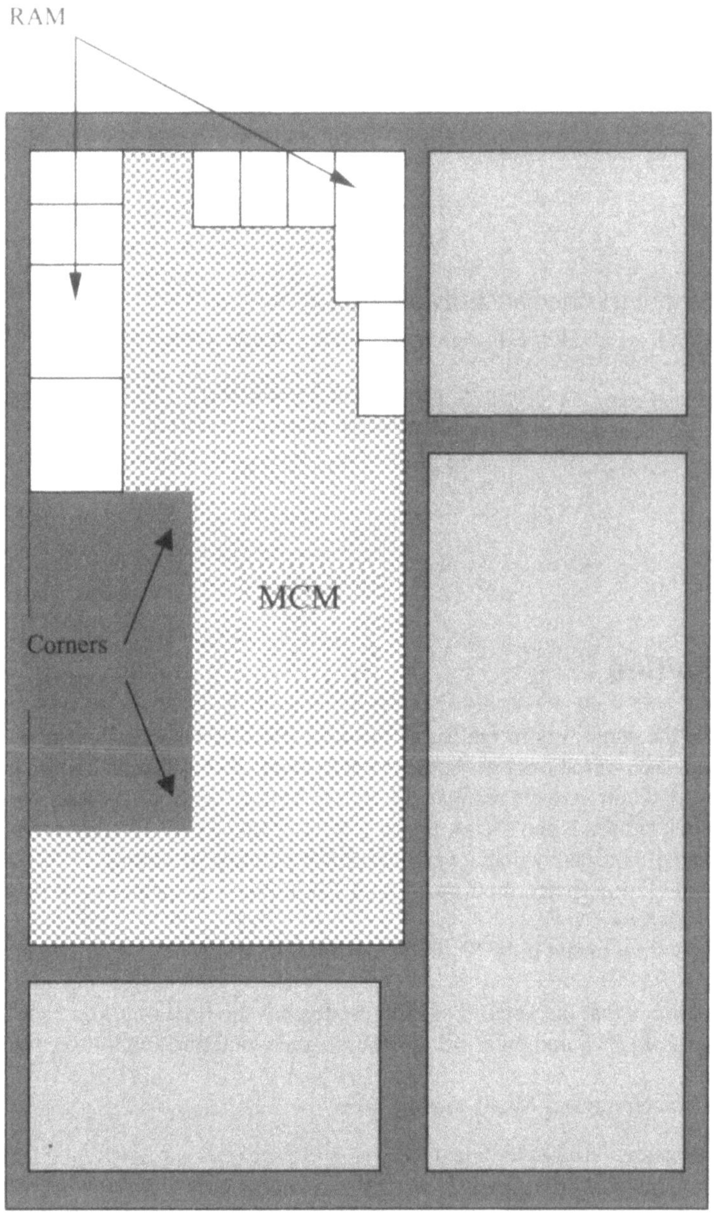

Figure 1. Integrated circuit top view

This is more evident looking at the block diagram of Figure 2:

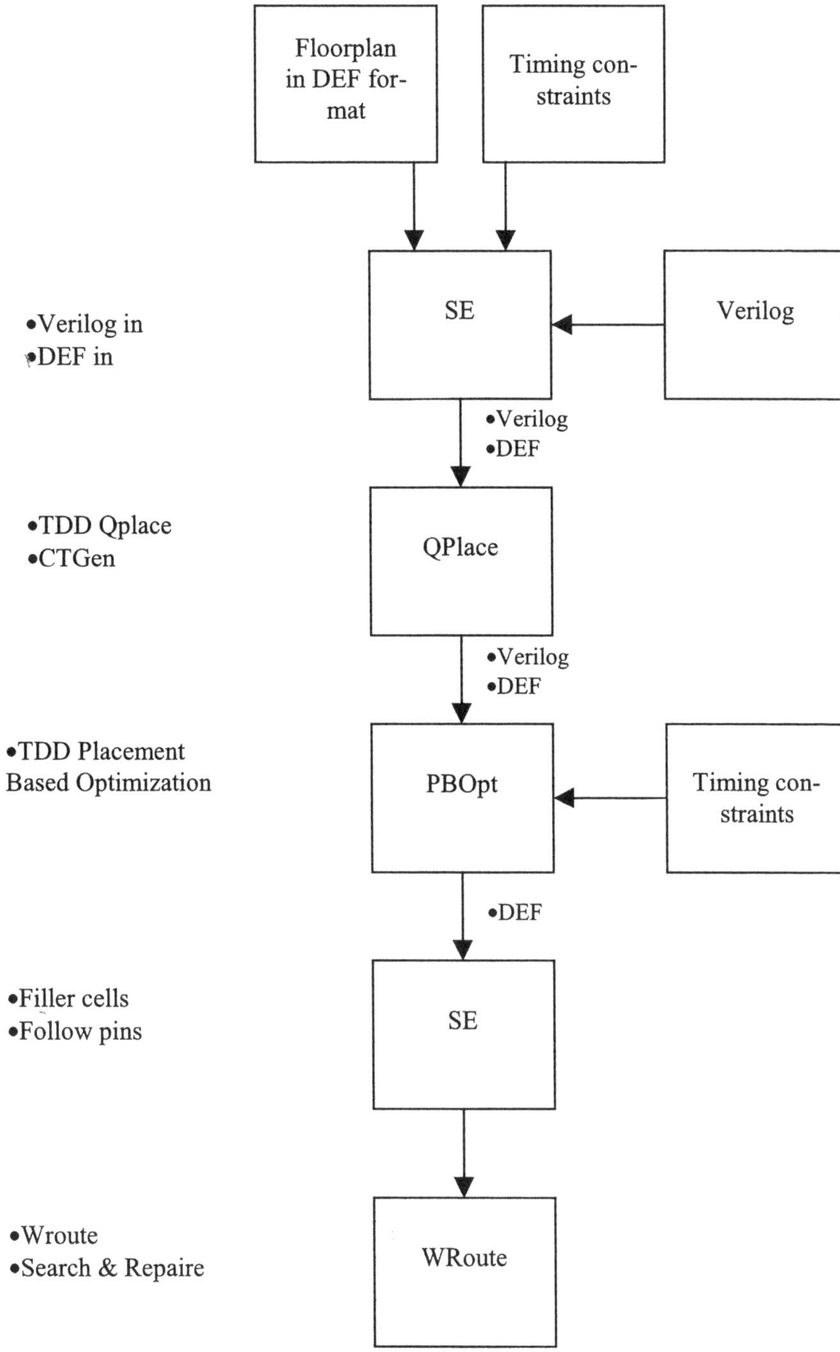

Figure 2. SE Design Flow

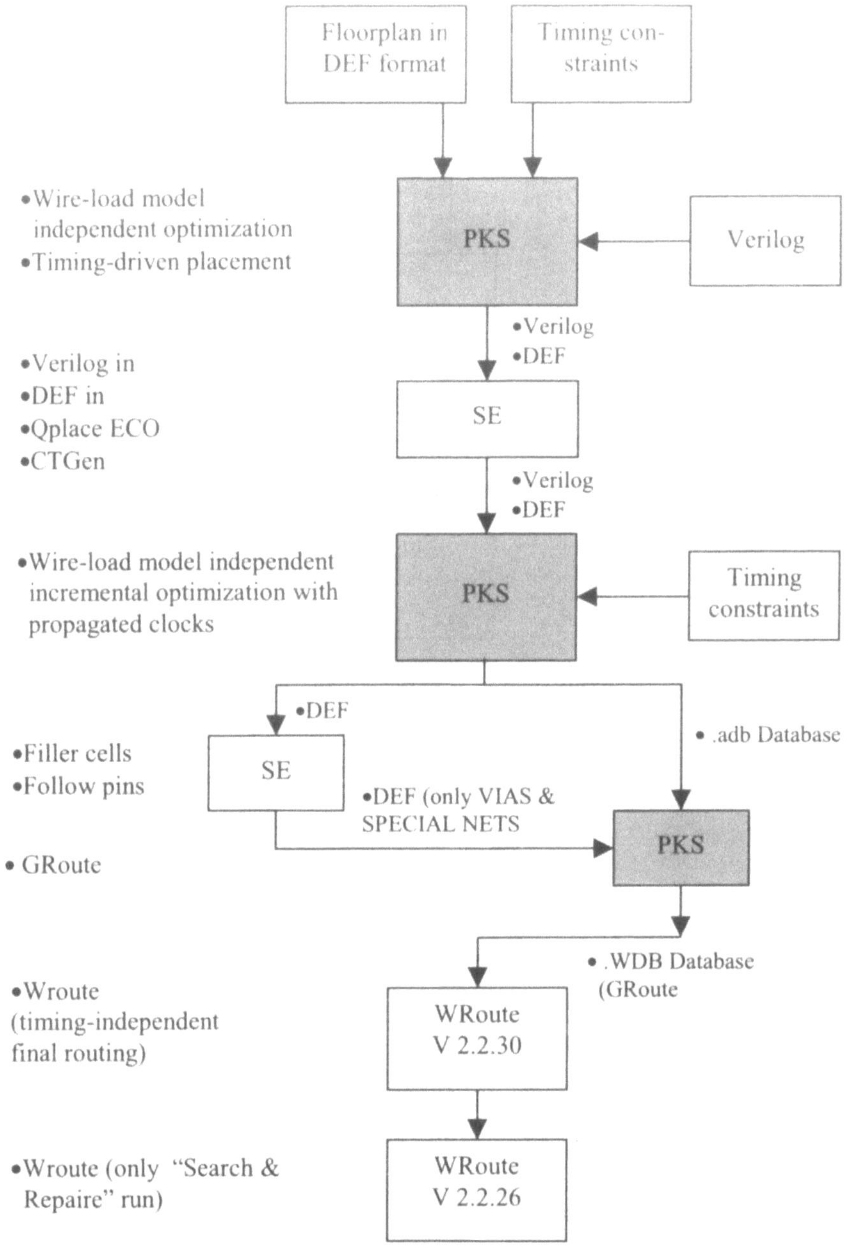

Figure 3. PKS Design Flow

NOTE: It is necessary to import Verilog together with DEF file because PKS can extract from DEF files only placement and obstruction information (*i.e.*, NET section is ignored).

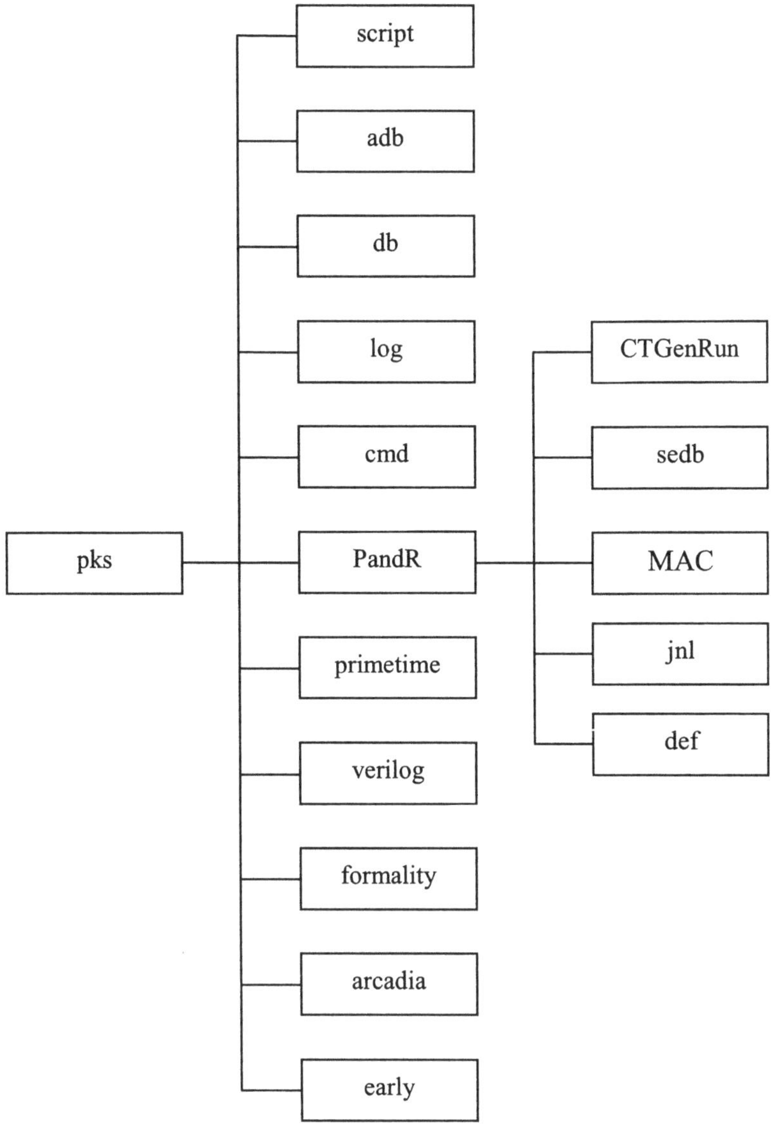

Figure 4. Directory structure

Directories

PKS tool needs the directory structure shown in Figure 4, to correctly perform its task.

In /PandR/sedb/ directory are saved SE data base, in /PandR/jnl/ files.jnl where is reported step by step what has been done by SE.

Analagously, PKS steps in files.log inside /log/ directory are reported.

In /script/ must be put the scripts written by the customer that allow automatic execution of a set of commands.

In /PandR/MAC/ directory are stored files written by the customer and invoked by instructions inside other script.

The same thing can be said for files.cmd in /cmd/ directory.

Script to Run PKS Flow

Hereby is reported a macro to run automatically a complete PKS Place and Route flow.

All characters after "#" are ignored.

```
########################################################################
#       MACRO FOR PKS
########################################################################

########################################################################
#       OPTIMIZATION
########################################################################

pks_shell    -visual    -f    script/MCM_DEMOD_pks_opt.tcl    -logfile
log/MCM_pks_over.log-cmdfile cmd/MCM_pks_opt_over.cmd

########################################################################
#       TRANSFER OF FILES TO SE
########################################################################

pks_shell  -visual      -fscript/to_SE.tcl-logfile log/to_SE.log
-cmdfile cmd/to_SE.cmd

########################################################################
#       SE
########################################################################

seultra  -m=400  -j=jnl/vin.jnl  -gd=ansi  "EXE MAC/verilog_in.mac;
quit; "

seultra  -m=800  -j=jnl/qplace_eco.jnl  -gd=ansi
"EXE MAC/qplace_eco.mac; quit; "

ctgentool ctgen cmd

pbodiff  -pbopt  -cmd  ./MAC/pbodiff.cmd

pbohvo  -pbopt  -cmd  ./MAC/pbohvo.cmd
```

```
#########################################################################
#        INCREMENTAL OPTIMIZATION
#        WITH PROPAGATED CLOCKS
#########################################################################

pks_shell -visual -f script/opt_incr_clock_prop.tcl -logfile
log/opt_incr_clock_prop.log -cmdfile cmd/opt_incr_clock_prop.cmd

#########################################################################
#        FOLLOW PINS AND FILLER CELLS GENERATION
#########################################################################

seultra -m=800 -j=jnl/placed2filled.jnl -gd=ansi "EXE MAC/
placed2filled.mac; fquit; "

#########################################################################
#        GLOBAL ROUTING
#########################################################################

pks_shell -visual -f script/global_routing.tcl -logfile log/ global_routing.log
-cmdfile cmd/global_routing.cmd

#########################################################################
#        FINAL ROUTING (WROUTE)
#########################################################################

wroute.V.2.2.30 -q MAC/final_routing.tcl -l ./jnl/final_routing.jnl

#########################################################################
#        POST ROUTE CORRECTION
#########################################################################

wroute -q MAC/wroute_fix.cfg -l ./jnl/ wroute_fix.cfg

#########################################################################
```

Comparisons

Data obtained in each step with PKS (new flow) are compared with those obtained with SE (standard flow).

After Qplace, that is first placement, the data related to the length of wire, number of instances and average row utilization obtained, that are compared to those obtained with PKS. The reports are shown Table 1:

Table 1. Results after Qplace

	PKS FLOW	STANDARD FLOW
Estimated wire length	13,420,345 μm	12,372,571 μm
Total wire length	13,420,345 μm	11,843,246 μm
Total weighted	16,437,354 μm	14,486,237 μm
Number of instances	81,937	78,665
Average row utilization	56.36%	56.59%

The length values are expressed in micron. From these data it is possible to see that the wire in PKS flow are greater than in standard one, and this is due to the fact that in this first step the PKS tool achieves a first optimization, that causes the instances to increase in number, but to decrease in size.

This can be noted, by looking that the average row utilization has not changed.

After CTGen step (Clock-Tree Generation), the results obtained are listed in Table 2:

Table 2. Results after CTGen

	PKS FLOW	STANDARD FLOW
Max skew	0.367 ns	0.369 ns
Number of components	288	605

It can be seen that the max skew between tree leafs is almost the same in both flows.

The number of components added by clock-tree generation tool (CTGen), in PKS flow, is little compared to that added in standard flow.

This is, obviously, a great advantage.

The next step was optimization, and Table 3 shows what has been obtained:

Table 3. Results after optimization

	PKS FLOW	STANDARD FLOW
Estimated wire length	13,749,359 μm	12,300,125 μm
Total wire length	13,761,360 μm	12,300,991 μm
Total weighted wire length	16,948,845 μm	15,216,903 μm
Average row utilization	56.53%	57.51%
Number of instance	82,226	81,721
Number of instance after Qplace	81,937	78,665
Instance added	289	3,056
Percentage of instance added	0.35%	3.88%

It can be seen that the wire length is greater in PKS flow than in standard one, while average row utilization improved, and number of instances added after optimization is much smaller.

In fact in new flow we have only 0.35% of instances added to those placed by Qplace, while in standard one this value is 3.88%.

After last step (Wroute), total wire length can be read:

Table 4. Results after Wroute

	PKS FLOW	STANDARD FLOW
Total wire length	13,767,799 μm	15,710,429 μm
Percentage of wire saved	12.36%	

Remarkable improvement in the wire length has been registered.

Conclusion

All parameters of our interest are better with PKS tool than those obtained with SE tool (standard flow).

The drawback is a major complexity in tool usage, but it should be remarked that the one used, is a beta release. The next release should be simpler as it should avoid to exit from PKS and enter SE to run some steps.

Battery Modelling In Embedded Systems

Maria Ausilia Gemma
Donata Nicolosi

STMicroelectronics S.r.l.
Stradale Primosole 50, 95100 Catania – Italy

e-mail: mariaausilia.gemma@st.com
donata.nicolosi@st.com

Introduction

Progress in battery technology is closely tied to that in electronics. The fast development in battery-powered portable systems and the increasing demand for longer run time and lighter weight handheld devices is driving battery makers to make new investments and researches into new battery technologies. This resulted in a great improvement in energy density, shelf life and reliability of batteries. However the battery technology progresses are very slow compared to the ones in the electronic field. A typical battery-powered System is represented in Figure 1 [1]. This system is present in portable devices such as laptop computers, cellular phones, *etc*. It consists of the VLSI circuit, the dc-dc converter and the battery.

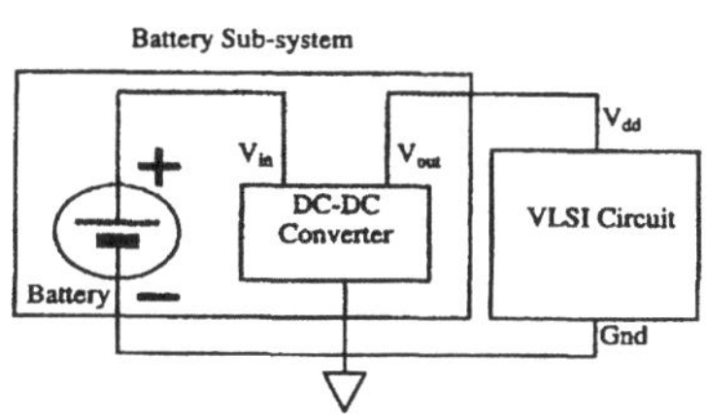

Figure 1. A complete battery powered system

The design of future portable systems requires high performance of the batteries with low thermal dissipation and long life. The average energy consumption is directly related to the battery life; hence it may be the critical factor that sets the system weight and its cost.

Different battery systems have been developed in the last years to meet demands for increased energy density and longer life: among these, the Li-Ion and the NiMH are the most common for portable systems.

There are two philosophies in selecting a battery system, each addressing a given market sector. The commercial market is looking for minimum size but maximum energy density for long run-times; the industrial market focuses on reliability and extended service life.

Batteries in embedded systems are designed taking into account two different levels:

- physical and mechanical level: the main targets at this level are high power, capacity, reliability and lightness;
- electronic level: the main target at this level is the optimization of using electrical and electronics resources.

So far these two levels have been always considered separately but now we propose to consider both together in order to reach an overall optimization of the system. The figure 2 shows a logical scheme of the framework we suppose to adopt to design very efficient portable systems.

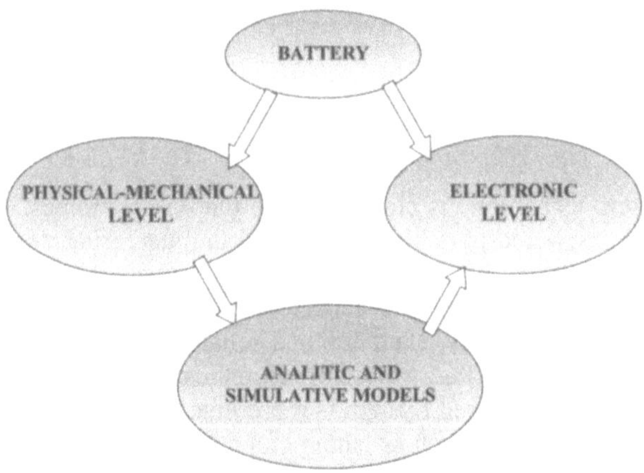

Figure 2. Framework for designing very efficient portable systems

We developed analytic and simulative models of the physical behavior of the batteries and these models can be practically used at the electronic level.

The advances of portable systems require tight constraints on energy consumption. The energy delivered by a battery depends upon the rate at which the power is withdrawn: in general, if the battery current increases, the amount of energy delivered decreases [2]. The energy depends also on the battery construction, by the cell size and by the duty cycle employed [3]. The rapid progress in semiconductor process technology caused the chip density and the clock frequency to increase. Thus, the power consumption in digital circuits became a critical issue: the higher the power consumption, the smaller the battery life.

Usually the power reduction is obtained by hardware techniques, such as varying the clock frequencies and the supply voltages. However, hardware devices consume power to serve requests from software programs. Hence, it is essential to consider software for power reduction. An energy-efficient design of battery-powered embedded systems requires the optimization of both hardware and software.

The main target of our work has been the development of the battery's physical model and the creation of its analytic discrete model in order to use it to simulate batteries behavior.

Battery Technology

A battery is a device that converts the chemical energy contained in its active materials directly into electric energy by means of an electrochemical oxidation-reduction (redox) reaction. This type of reaction involves the transfer of electrons from one material to another through an electric circuit. A battery consists of one or more of these cells, connected in series or parallel, or both, depending on the desired output voltage and capacity. The cell consists of three major components, as shown schematically in Figure 3 [2]:

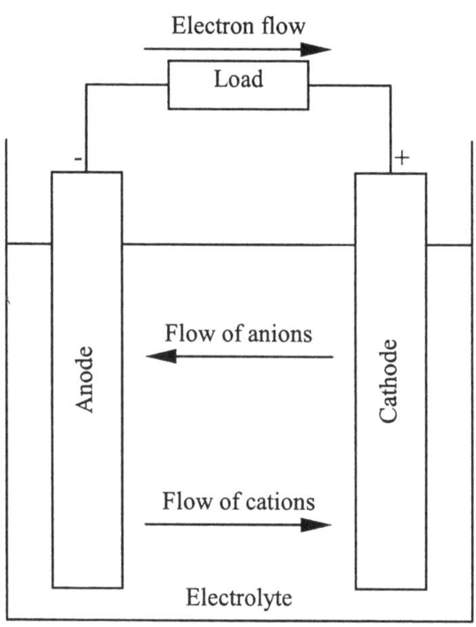

Figure 3. Electrochemical operation of a cell (discharge)

- the *anode* or *negative electrode* – the reducing or fuel electrode – which gives up electrons to the external circuit and is oxidized during the electrochemical reaction;
- the *cathode* or *positive electrode* – the oxidizing electrode – which accepts electrons from the external circuit and is reduced during the electrochemical reaction;

- the *electrolyte* – the ionic conductor – which provides the medium for transfer of electrons, as ions, inside the cell between the anode and the cathode. Usually, the electrolyte is a liquid, such as water or other solvents, with dissolved salts, acids, or alkalis to impart ionic conductivity. Some batteries use solid electrolytes, which are ionic conductors at the operating temperature of the cell.

The fundamental terms that define the characteristics of the battery are the following:

- *CAPACITY:* ampere-hours available from a fully charged cell or battery.
- *CAPACITY, AVAILABLE:* total capacity that may be obtained at defined charge and discharge rates and their associated environmental conditions.
- *C RATE:* discharge or charge current rate in ampere, numerically equal to rated capacity of a cell in ampere-hours, rated at the 1 h discharge rate.
- *CUT-OFF VOLTAGE:* voltage at which a discharge or charge is terminated.
- *C/X DISCHARGE RATE:* the current that would completely discharge a fully charged battery in x hours. For example, a battery with a total capacity of 90 ampere-hours has C/3 discharge rate of 30A.
- *DISCHARGE VOLTAGE:* the voltage of a battery during discharge.
- *RATED CAPACITY:* a designation by the battery manufacturer which helps identify a particular cell model and also provides an approximation of capacity; usually is expressed in ampere-hours at a given discharge current.

The development of new portable applications, such as computers, with high power requirements, accelerated the use of rechargeable batteries (secondary batteries) because of their power advantage over conventional types of primary batteries (not rechargeable batteries).In this work we consider only secondary batteries.

A rechargeable battery gradually looses charge because of aging and cycling. This wear-down effect varies with battery chemistry; some systems endure vigorous cycling, others are built for maximum energy density. Charge method, depth of discharge (DOD), ambient conditions and maintenance procedures also influence the service life of a battery. These factors have to be taken into account when creating a battery model.

The Lithium Battery

Initially, nickel-cadmium (NiCd) batteries were the only rechargeable batteries available for decades. In the 1980s, battery manufacturers launched a new technology – nickel-metal hydride (NiMH) – that boosted capacity. In the 1990s, the Sony Corporation introduced in the market the lithium-ion (Li-ion) batteries.

At the beginning, in Li-ion batteries the anode was made of lithium metal but these cells were characterized by lithium deposits that increased the cell impedance, limiting the cycle life, and created safety problems. In modern Li-ion batteries, the anode is replaced with a carbon material (Coke or Graphite) which stores electrons and lithium ions during the charging period and gives them up during the discharge period [3].

The cathode consists of an oxide that stores electrons and lithium ions during the discharging period and gives them up during the charging period. No lithium metal is involved in the reaction but only ions and electrons so there aren't safety and environmental problems. Thus, Li-ion batteries offer advantages that neither the NiCd nor the NiMH can meet: higher energy density and longer runtime are some of their advantages. Moreover, since the Li-Ion cells have an operating voltage that is three times bigger than both NiMH and NiCd, fewer cells are required for each battery pack, and the weight and the size are remarkably reduced [2]. Thus, because of the need for more power, many batteries suppliers have placed big hopes in lithium-ion (Li-ion) technology to power many of the new emerging portable devices, so, Li-ion batteries have become the most popular batteries in portable devices.

On the other hand it must be observed that the numerous advantages of Li-ion battery chemistries over nickel-based chemistries (NiCd and NiMH) are offset by the additional care and diligence required to ensure Li-ion charging system accuracy and safety.

Battery Models in Scientific Literature

Circuit-level, continuous-time battery models [4], [5], [6], [7] have been developed to help designers estimating the discharge characteristics of common batteries; such characteristics can be measured by connecting the actual battery to a system prototype.

In the scientific *literature* we have found four different developments of battery models, which utilize different approaches. All these approaches have as their main goal, the battery life-time optimization.

Spice Battery Model (S. Gold)

This approach presents a parametric PSPICE macromodel for simulating Lithium-Ion batteries. Figure 4 shows this model.

The cell voltage, state of charge (SOC), rate dependence, and thermal characteristic of the battery are all modelled in separate loops. The currents and voltages in these loops are analogs of physical variables within the cell. The current flowing through the model's *plus* and *minus* terminals is measured and used to calculate a normalized state of charge, which in turn is used to determine the cell voltage [7].

The state of charge in the model is represented as the voltage across a capacitor (C_CAP) with corrections for rate dependence and usage. The initial value of C_CAP is calculated based on the rated capacity:

$$C\_CAP = 3600\ (s/h)\ CAPACITY\ (Ah)$$

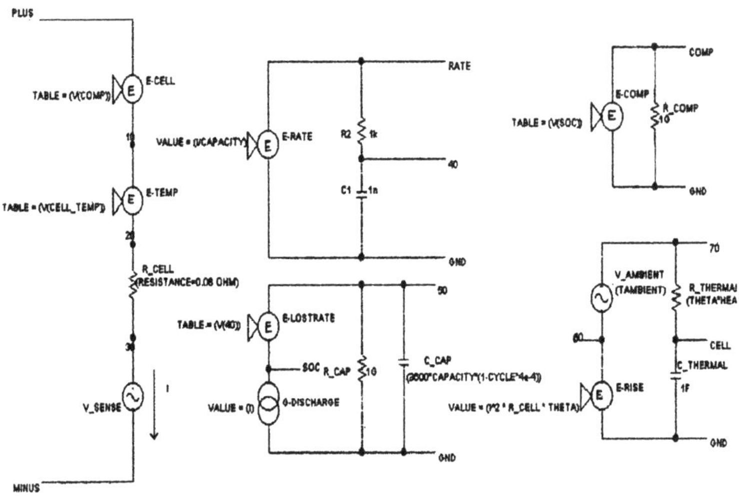

Figure 4. Lithium-ion macromodel schematic

The voltage across the cell becomes the ratio of transferred charge to maximum cell capacity.

$$V(C\_CAP) = Q/C = \frac{\int Idt}{C}$$

Changes to the state of charge are made by sensing the battery current with a zero voltage source, V_SENSE, and applying that current to C_CAP via the current-controlled current source G_DISCHARGE. The rate dependence of capacity is modelled with a voltage (E_LOSTRATE) that subtracts from the voltage across C_CAP. The ratio of battery current to capacity is the basis for changing the state of charge. This ratio sets the voltage at the RATE node, which becomes the index into a table containing the rate dependence on capacity. A single RC time constant models electrochemical delays in the battery that prevent instantaneous changes in cell voltage.

$$\tau = R\_RATE \cdot C\_RATE \cong 1 \text{ second}$$

The normalized state of charge is converted into the cell voltage via the measurement data contained in E_CELL generator.

The modelling technique described is quite general and there are some dynamics in the cell that are not in the model.

VHDL Battery Model (Benini-Castelli-Macii-Poncino-Scarsi)

The complete discrete-time model is implemented as a VHDL entity that well approximates the continuous-time behavior but, at the same time, it is fast and efficient enough to enable high-level, event-driven simulation of a complete system description.

Charge storage in a battery can be modelled as a capacitor with a capacitance C = 3600 · CAPACITY, where CAPACITY is the nominal capacity in Ahr, which is usually provided in the battery's data-sheet [8], [9], [10]. By setting the initial voltage across the capacitor $V_c=1$, they initialized the battery to its fully charge state. The simple linear capacitor model is not accurate enough to model complex phenomena observed during battery discharge.

They modelled the battery behaviour at the circuit level with the circuit in Figure 5:

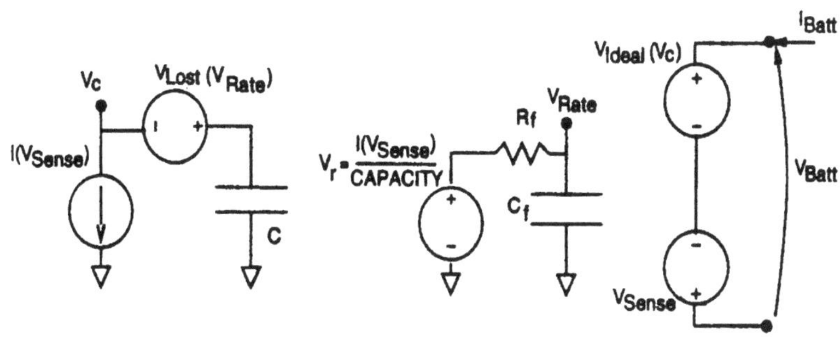

Figure 5. Continuous-time battery model

Dependency on the SOC is taken into consideration by storing several points of the curve in a look-up table (LUT) addressed by the value of the state of charge (V_c). Dependency on discharge rate is modelled with a voltage source V_{lost} in series with the charge storage capacitor. Voltage V_{lost} reduces the apparent charge of the battery (which controls output voltage). The value of V_{lost} is a non-linear function of the discharge rate (which can be modelled by another LUT). Dependency on the discharge frequency is modelled by averaging the discharge rate used to control V_{lost} through a low-pass filter (R_f, C_f). Notice that the voltage V_r in Figure 5 has to be multiplied by a correcting factor K.

The internal structure of the battery is based on the circuit-level model of Figure 5, and it consists of two concurrent, communicating processes. The first one computes the value of node V_c in Figure 5, the instantaneous state of charge of the battery. The second process computes the value of V_{lost}, it implements the low-pass filter shown in Figure 5. The output voltage of the battery, V_{in}, is a function of V_c. It is implemented in VHDL with a continuous assignment: $V_{in}=F(V_c)$, where F is realized by a LUT with linear interpolation.

In this model the recovery effect is evaluated through the time constant (R_f, C_f).

Stochastic Battery Model (Chiasserini-Rao)

In this model the theoretical capacity of the battery is defined as the maximum number of packets that the battery is able to deliver for a given value of current. Assuming that the amount of capacity necessary to transmit a packet is one charge unit, they set the theoretical capacity equal to T charge units and the initial value of charge of the battery to N charge units. N is also assumed to be the number of charge units that can be drained from the battery under constant discharge [11], [12]. The battery behavior is represented as a discrete time Markov process with initial state equal to N and one absorbing state, 0, corresponding to complete discharge, such as shown in Figure 6:

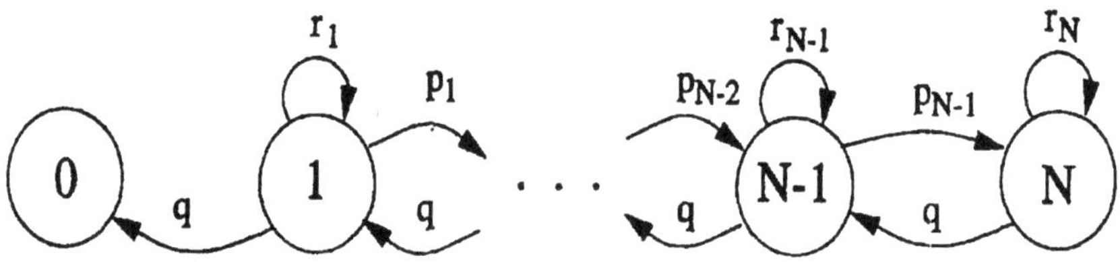

Figure 6. Markov chain representing the battery behavior

In case of constant current discharge the battery transmits N successive packets and goes from N to 0 in N time units.

In this model the recovery effect is represented as a decreasing exponential function of the state of charge of the battery. To more accurately model real cell behavior, the exponential decay coefficient is assumed to take different values as a function of discharged capacity [11], [13], [14].

Analytical Battery Model (Šimunić-Benini-De Micheli)

This battery model has been integrated with a simulation environment to provide cycle-accurate estimates of battery life. The main battery characteristic is its rated capacity. The battery model computes the battery efficiency losses due to the difference between the rated current and the average discharge current [15], evaluating the battery energy loss per cycle and the battery life-time.

As battery cannot respond to instantaneous changes in current, a first order time constant τ is defined to determine the short-term average current drawn from the battery [7]. Given τ, and processor cycle time T_{cycle}, the number of cycles N_{bat} can be computed:

$$N_{bat} = \frac{\tau}{T_{Cycle}}$$

then, the average current is calculated as follows:

$$I_{ave} = \frac{1}{N_{bat}} \sum_{cycle=1}^{N_{bat}} I_{system}(cycle)$$

The discharge rate is given by:

$$R_I = \frac{I_{ave}}{I_{rated}}$$

where I_{rated}, the rated discharge current, is derived from the battery specification and I_{ave} is the average current requested by the load [15].

Efficiency is calculated, given the discharge rate, using linear interpolation between the points from the discharge plot in Figure 7.

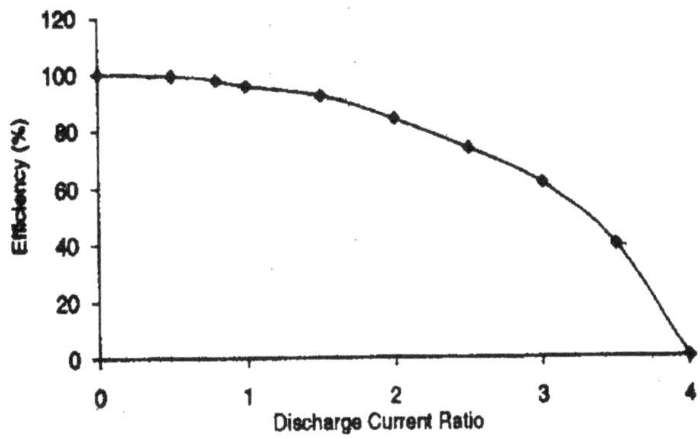

Figure 7. Battery efficiency

Given the efficiency loss (1-η), the total battery energy loss per cycle is evaluated:

$$E_{bat} = (1-\eta)\, I_{ave}\, V_{Bat}\, T_{cycle}$$

Lower efficiency means that less battery energy remains and therefore the battery life-time is proportionally lower.

Analytical Battery Model

To develop the battery model, we considered the main effects that characterize the battery behavior.

Batteries are electrochemical devices that convert chemical energy into electrical energy by a redox reaction. Obviously, it would be desiderable, if, during the discharge, all the theoretical energy could be converted into electric energy. However, energy limitations exist and the maximum electric energy that a battery can deliver depends on many factors. When a battery is discharged, its voltage is lower than the theoretical voltage, this is due to two important factors:

- the polarization effect, that consumes part of the energy which is given off as waste heat;
- the internal impedance of the cell, that causes a voltage drop (IR loss) during operation, which also consumes part of the energy as waste heat.

The battery voltage strongly depends on the current load (discharge rate): as the load is increased, the IR loss and then the voltage drop increases; so, the battery service life is reduced. Batteries datasheets show the effect of discharge rate on battery energy density and battery capacity. As the discharge rate increases the coulombs delivered decreases as well as the energy density. If the current drawn from the battery increases, the duration of the discharge to the point where the end-point voltage is reached decreases, and the voltage profile decreases: this means that with rapid discharges lower voltages can be obtained. The voltage profile at different battery discharge rate is shown in Figure 8: as the rated capacity increases, the voltage decreases till it reaches the cut-off voltage of 2.7 V at which the battery is fully discharged. Figure 8 shows different curves, each of them referred to a different load I:

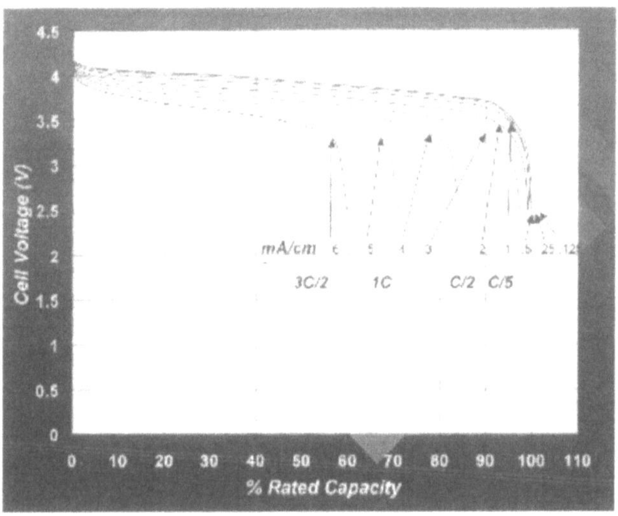

Figure 8. Voltage-profile plot (courtesy of Valence)

If the battery is subjected to a pulsed load, the voltage drop depends also on the pulse length: operating at two or more discharge load is typical of electronics devices as Camcorders, Notebook and Cellular Phones, because of the different functions they perform.

Previous researches have shown that the specific energy (delivered energy/weight of the cell) delivered by a battery cell can be increased by keeping the drained current below a specified rated discharge current and by inserting idle time periods between successive battery discharges. During the idle period, certain chemical and physical changes take place and the result is a Voltage growth. This phenomenon is known as *Recovery Effect*.

Battery Discharging Model

We developed a discrete time-varying battery model, in order to write the C algorithm easily. To develop the model we used the following variables:

- I_n is the current drawn from the battery by the system.
- V_n is the battery voltage when a current load I_n is required.
- Δt is the microprocessor cycle time.
- Q is the battery nominal capacity (initial capacity).
- Q_n is the battery capacity at time *n*.

Since, the battery capacity is strongly dependent on the discharge current I_n, we defined the capacity at time *n*+1 as:

$$Q_{n+1} = Q_n - I_n \Delta t / \eta (I_n)$$

where

$$I_n \Delta t / \eta (I_n)$$

is the capacity used during the time period Δt, and

$$\eta (I_n) = Q_n / Q$$

is the battery efficiency or utilization.

The battery utilization η is a function of the current I_n. Figure 9 shows the battery efficiency variation versus the discharge rate.

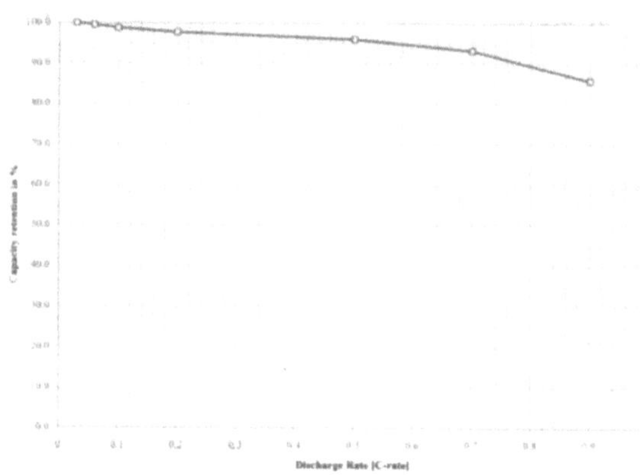

Figure 9. Capacity retention vs. discharge rate at 23°C (courtesy of Valence)

Battery efficiency is calculated using linear interpolation between the points from the discharge plot in Figure 9. The voltage depends on the battery capacity:

$$V^*_{n+1} = K (Q_{n+1})$$

This dependency is taken into consideration by storing several points of the curve shown in Figure 8, in a look-up table (LUT) addressed by the value of the state of charge (Q_{n+1}). K is a function of Q/V and is calculated from the LUT, using linear interpolation. During the time Δt, at which the load is connected to the battery, this is discharged with a time constant τ_1; thus, the voltage at time $n+1$ is:

$$V_{n+1} = V^{*}_{n+1} - \frac{\Delta t}{\tau_1}\left(V^{*}_{n+1} - V_n\right)$$

The time constant τ_1 models the battery discharge when the battery is connected to the load.

Battery Recovery Model

We suppose that, if at time t_0 the current I_n drawn by the load is below a threshold value, after a time period t_{rest}, the recovery effect takes place. The recovery effect is characterized by an exponential growth that is modelled with the time constant τ_2. The equation that describes the exponential charging from one initial value V_0 at time t_0 to a final value V_1 at time $t_0 + \Delta$ is:

$$V(t) = V_0 + (V_1 - V_0)\left(1 - e^{-\frac{t-(t_0+\Delta)}{\tau}}\right)$$

Discretizing and approximating the exponential with its Taylor series arrested to the first order:

$$e^{-\frac{\Delta t}{\tau}} \cong 1 - \frac{\Delta t}{\tau}$$

we obtain:

$$V_{n+1} = V_n + (V_1 - V_0)\left[\frac{t_{n+1} - t_n}{\tau 2}\right]$$

then:

$$V_{n+1} = V_n + \frac{\Delta t}{\tau 2}(V_1 - V_0)$$

Let's say:

$$V_{last} = V_1; \quad V_n = V_0; \quad \tau_2 = \tau$$

the recovery discrete model is:

$$V_{n+1} = V_n + \frac{\Delta t}{\tau_2}(V_{last} - V_n)$$

where V_{Last} is the value of the nominal voltage.

Battery estimation is performed as follows. Given the data taken from the data sheets, the battery is characterized with its capacity, its voltage, the two time constants and the two table of points. The first table describes the battery efficiency plot at different discharge rates and the second describes the battery's voltage profile (voltage characteristics versus the rated capacity).

During each simulation cycle Δt, a constant discharge current is considered. Given the discharge current the battery's voltage is computed in the following way: if the current value is bigger than a fixed threshold value, the discharge rate is computed; then, using linear interpolation between the points of the efficiency plot, the battery efficiency is computed. Given the efficiency, the new battery capacity and the rated capacity are computed. Given the rated capacity, using linear interpolation between the points of the voltage profile plot, the voltage is computed. Finally, the battery voltage is computed considering a time constant of 1s and then the battery energy is evaluated.

If the discharge current value goes under the threshold value, the recovery effect takes place: after waiting for a small rest period the battery recovery effect starts and the battery voltage increases with a time constant τ_2=150s.

Simulations C

Many simulations have been run to verify the simulator good behavior. For this reason, in order to minimize the running time, for most of them we used a clock of 1second with a discharge time constant τ_1= 10 seconds and a recovery time constant τ_2=1500 seconds. To verify the battery-discharging model, a simulation at a discharge rate of 0.9 (C-rate=C/1.1=0.9C) has been taken into account. This means that a battery connected continuously to a constant load of 0.9C = 1.18 ampere will be fully discharged in 1 hour and 1 minute.

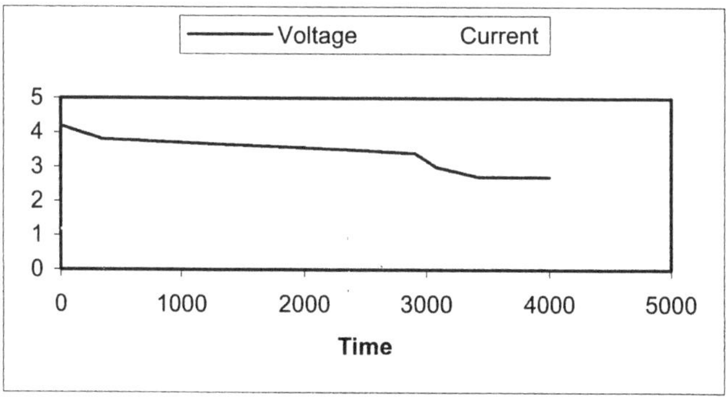

Figure 10. Battery discharge at C/1.1 rate

As shown by the plot in Figure 10, the battery voltage, starting from the nominal value of 4.2V, decreases till the cut-off value, at which the battery is considered fully discharged, in about 3660 seconds (that is in 1 hour and 1 minute).

The simulation has been repeated varying the load. A discharge rate of 0.2 (C-rate=C/5=0.2C) has been taken into account. This means that a battery connected continuously to a constant load of 0.2C= 0.26 ampere will be fully discharged in 5 hours. Two different input files have been used. The first input file uses a voltage-profile table whose values have been derived from the plot in Figure 8 . Figure 11 shows the battery discharge at a C/5 rate for the first input file.

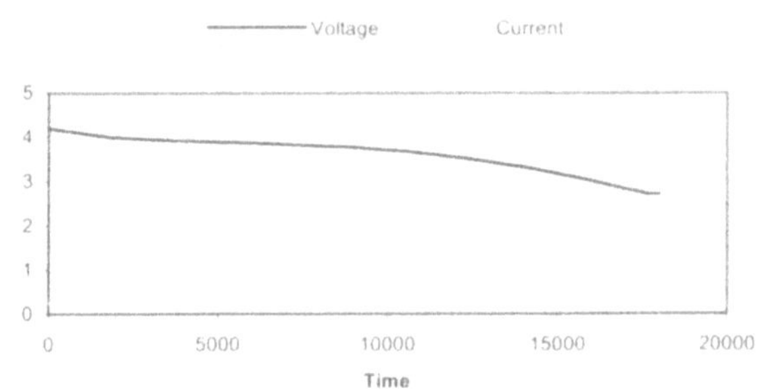

Figure 11. Battery discharge at C/5 rate, using the voltage-profile plot in Figure 8

The second input file uses a voltage-profile table for Polystor ICR - 18650 used in the S. Gold 's model [7].

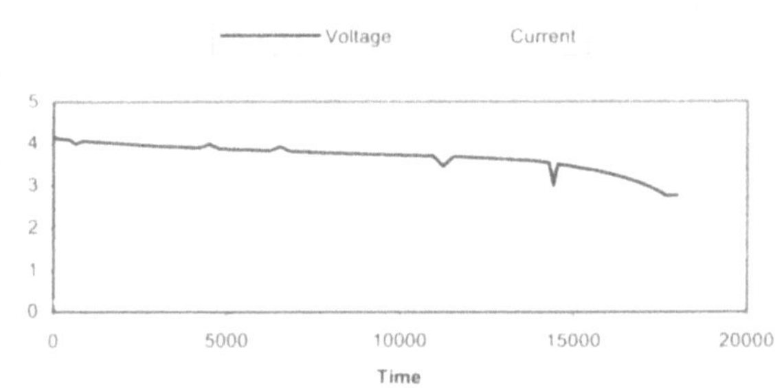

Figure 12. Battery discharge at C/5 rate, using data from [7]

In both cases, the battery reached the fully discharged state in 5 hours, as it was expected. The next step has been to study the battery discharging behavior, when a load step is applied. The simulation has been repeated varying the load. First, we considered a discharge rate of 0.9 (Figure 13).

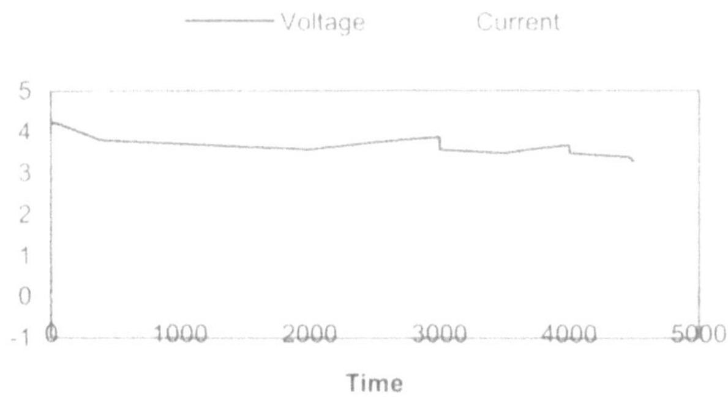

Figure 13. Battery discharge at C/1.1 rate, under a step load

During the rest period in which the battery is not connected to the load, we can see the presence of the recovery effect that makes the battery voltage to increase. Next, we computed the battery energy consumption.

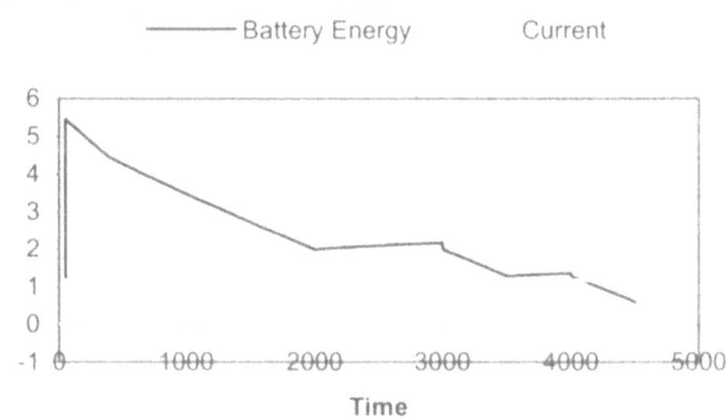

Figure 14. Energy delivered by the battery at C/1.1 rate, under a step load

Observing the energy plot in Figure 14, we can notice how the battery energy decreases during the period in which the battery is connected, but increases during the recovery phase. This is due to the fact that the even if the battery chemical energy decreases, the electrical energy increases with the voltage:

$$E = V \cdot C \quad [VAh]$$

The "Energy" in the energy plot represents the electric energy in ampere-hour delivered by the battery.

In the second simulation a discharge rate of 0.2 has been taken into account. This means that a battery connected continuously to a constant load of 0.2C = 0.26 ampere will be fully discharged in 5 hours. Once again, two different input files have been used. The first input file uses a voltage-profile table whose values have been derived from the plot in Figure 8.

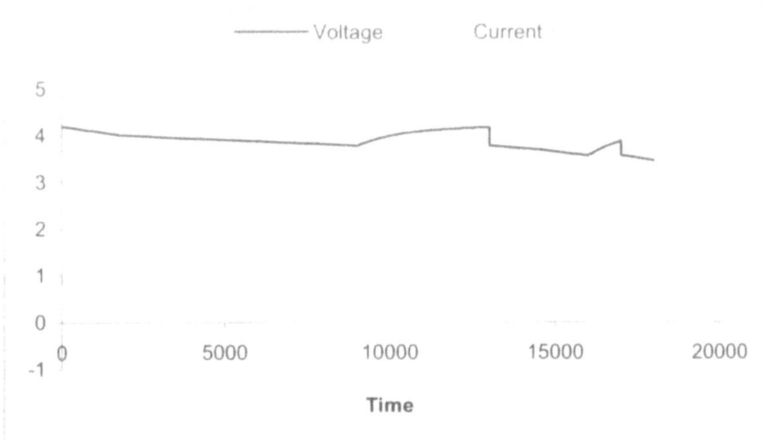

Figure 15. Battery discharge at C/5 rate, under a step load

The second input file uses a Voltage-Profile table for Polystor ICR - 18650 used in the S. Gold 's model [7].

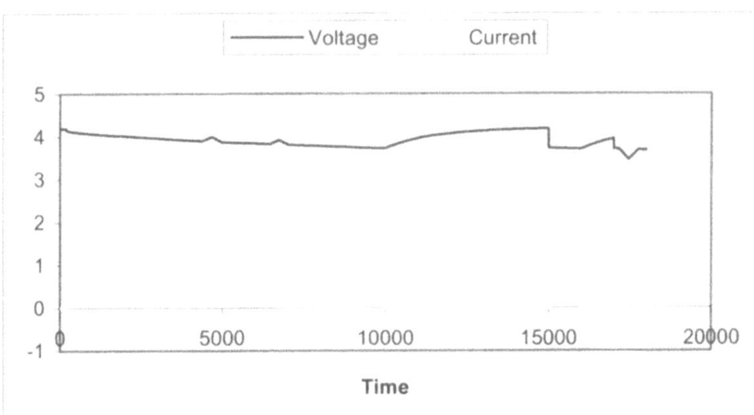

Figure 16. Battery discharge at C/5 rate, under a step load, using data from [7]

Again, during the rest period, it's possible to see the presence of the recovery effect, as shown in Figure 15 and in Figure 16. Obviously, in both cases the battery "Energy" decreases while the battery is connected to the load.

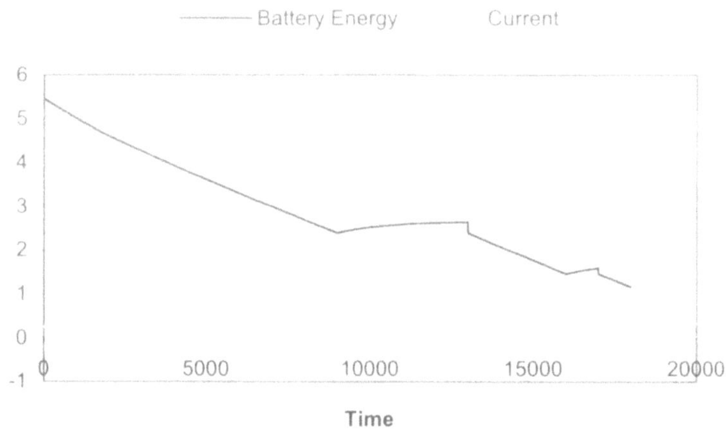

Figure 17. Energy delivered by the battery at C/5 rate, under a step load

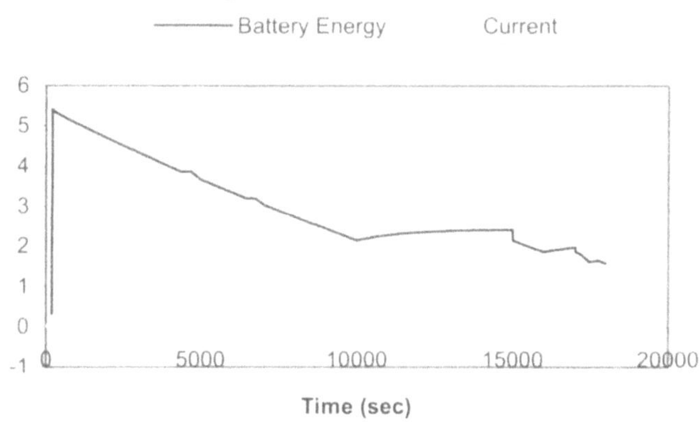

Figure 18. Energy delivered by the battery at C/5 rate, under a step load, using data from [7]

After running these simulations, we changed the clock cycle to 1msec using the same time constants as before. A discharge rate of 0.2C has been considered. The following plots show the voltage and energy behavior versus the time (in seconds), when the load varies.

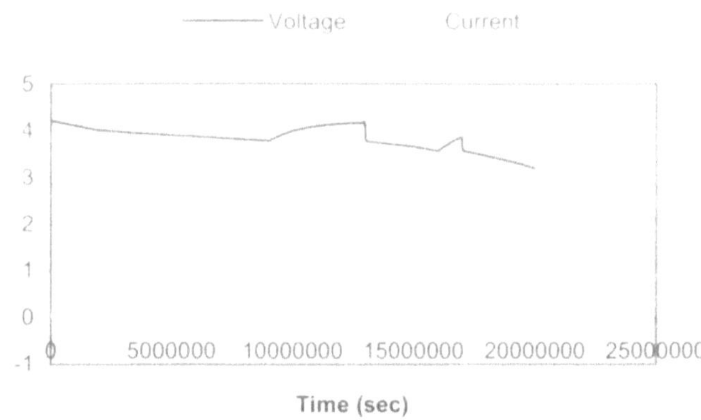

Figure 19. Battery discharge at C/5 rate, under a step load

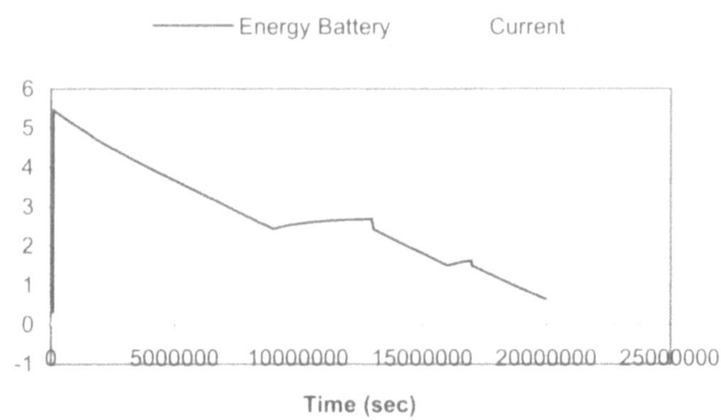

Figure 20. Energy delivered by the battery at C/5 rate, under a step load

The plots obtained are very similar to the ones obtained with the previous simulations (Figure 19), (Figure 20).

The CoWare Environment

CoWare N2C is an environment that helps system-on chip designers to create on silicon their ideas starting from its specifications. It provides designers the ability to perform true co-design and co-verification at different levels of abstraction during the design process. It helps eliminate long design cycles and, as such, it significantly reduces the time-to-market. It incorporates common hardware and software compilers and simulators and promotes reuse by encapsulating expert design knowledge, so that it can be reused by non specialists [16].

The major problem with today's system design is that it involves different teams, each one using different tools and methodologies but, to create a system on chip, digital, analog and mixed-signal hardware and software design have to be combined. In a traditional co-design approach, there is a gap between the hardware and software teams: the software team has to wait until the hardware has been designed, to start debugging its software in realistic conditions. Hardware and software design needs to be done sequentially. In this way errors made during the hardware and software design are discovered only during the hardware and software integration. Obviously, this increases the design process cost and introduces large delays in time-to-market. The CoWare design flow solves this problem.

CoWare provides three phases in the systems on silicon design process:

1. system specification;
2. partitioning and interface design;
3. hardware and software optimization.

In CoWare N2C there is a formal high-level specification which uses C. The system designer writes a functional specification that is implementation oriented but independent of the hardware architecture; when he is satisfied by the functional specification he gives it to the hardware/software partitioning team. Different hardware/software partitioning results are taken into account, the specification of the hardware/software part of the design can be handed over to the design teams. These teams can refine these specifications at different levels of abstraction and then can optimize the implementation. In this case the system specification doesn't change throughout the project and hardware/software design can be done in parallel. This improves considerably the time-to-market.

The system is made by blocks. To describe the implementation of a system's block two kind of encapsulations can be used:

- structure encapsulation: it allows to create a hierarchical decomposition of the system;
- host language encapsulation: it describes the behavior of a block.
- there are three types of host language encapsulations:
    - ✓ C encapsulation;
    - ✓ VHDL encapsulation;
    - ✓ VERILOG encapsulation;

We used a C encapsulation, which is C code encapsulated with an abstraction layer that provides concurrency and communication constructs. The encapsulation describes the behavior of a block using C language.

Simulations CoWare

To verify the battery-discharging model we run the simulations as in C. In order to minimize the running time, for most of them we used a clock of 1sec with a discharge time constant τ_1= 10 seconds and a recovery time constant τ_2= 1500seconds.

First, we considered a discharge rate of 0.9 (C-rate=C/1.1=0.9C).

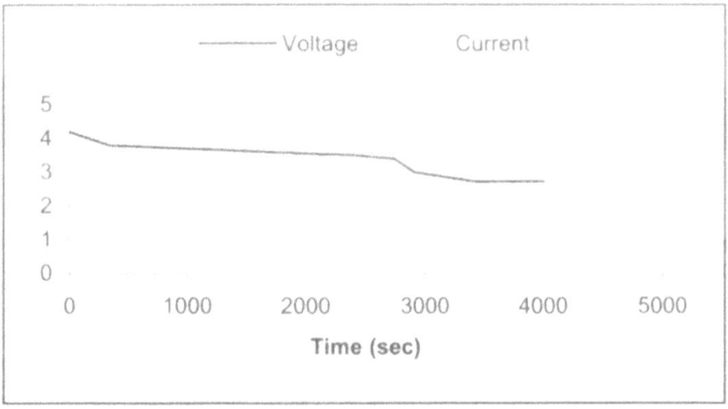

Figure 21. Battery discharge at C-rate =0.9

As it is possible to see from the plot in Figure 21, the battery voltage, starting from the nominal value of 4.2V, decreases till the cut-off value, in about 3660 seconds (that is in 1 hour and 1 minute). In the next simulation (Figure 22) we can see the battery behavior when it is subjected to a step pulse load.

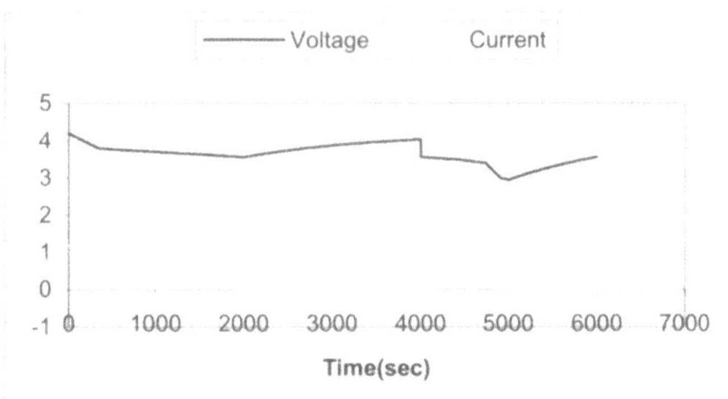

Figure 22. Battery discharge at C-rate = 0.9 (with recovery effect)

During the rest period in which the battery is not connected to the load, the recovery effect takes place.

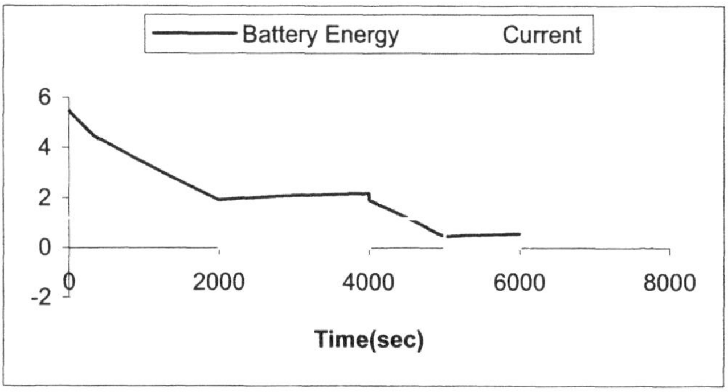

Figure 23. Battery energy at C-rate = 0.9

Figure 23 shows the battery energy behavior: as the battery is connected to the load the energy decreases. The simulation has been repeated varying the load. A discharge rate of 0.2 (C-rate=C/5=0.2C) has been taken into account. The battery has been connected to a step pulse load. Again, during the rest period, the voltage recovery takes place (Figure 24). Obviously, once again the battery "Energy" decreases while the battery is connected to the load, as it is shown in Figure 25.

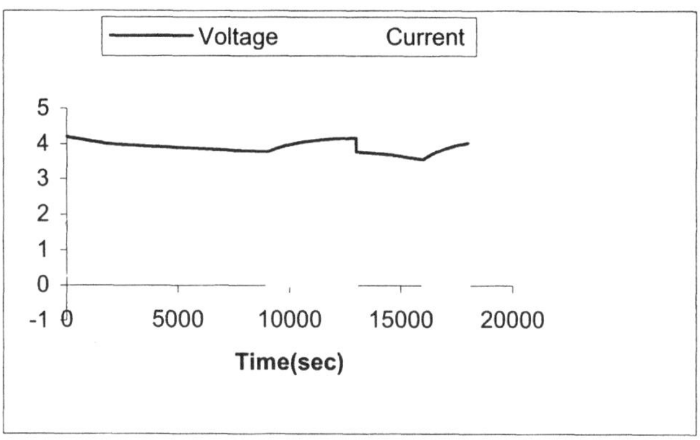

Figure 24. Battery discharge, at C-rate = 0.25, under a step load

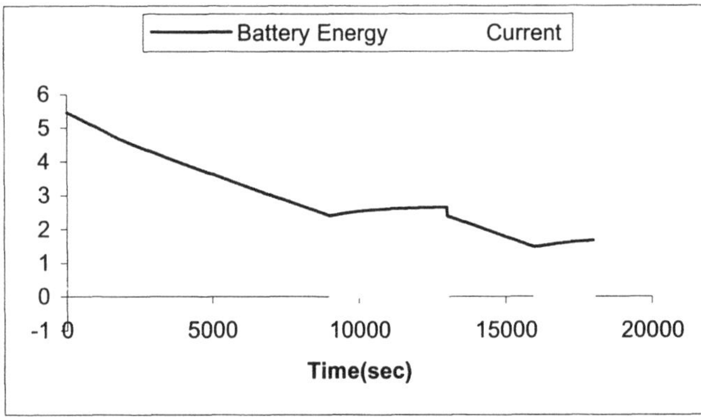

Figure 26. Battery energy at C-rate = 0.25

Conclusions

We developed a battery model simulator that considers the *recovery effect*. The Simulator is written in C code and encapsulated in the CoWare environment in order to have a more general system that could be used in different applications, and also to run realistic simulations by connecting realistic loads.

If the microprocessor clock increases but the voltage decreases, the energy decreases since the power is proportional to the squared voltage, but due to the batteries non-linear behavior this property has to be verified: the simulator could be used as a means to study the battery efficiency when the microprocessor voltage varies during the execution.

Future work will also include the development of a discrete model to evaluate dc-dc converter efficiency.

References

[1] M. Pedram and Q. Wu. Battery-Powered digital CMOS design, Proc. of Design Automation and Test in Europe. March 1999, 72-76.
[2] D. Linden. Handbook of Batteries and Fuel Cells. New York, McGraw-Hill.
[3] R. Powers. Batteries for low power electronics. Proc. of the IEEE, 83 (4):687-693, Apr. 1995.
[4] M. Doyle, T. F. Fuller, J. Newmann. Modelling of Galvanostatic Charge and Discharge of the Lithium/Polymer/Insertion Cell. Journal of the Electrochemical Society, 140 (6) 1526-1533, Jun. 1993.
[5] S. C. Hageman. Simple Pspice Models Let You Simulate Common Battery Types. EDN, 117-132, Oct. 1993.
[6] M. Glass. Battery Electro-Chemical Non-Linear/Dynamic Spice Model, Energy Conversion Engineering Conference. 292-297, Aug. 1996.

[7] S. Gold. A PSPICE Macromodel for Lithium-Ion Batteries. 12th Annual Battery Conference on Application and Advances, 215-222, Jan. 1997.
[8] L. Benini, G. Castelli, A. Macii, E. Macii, M. Poncino, R. Scarsi. A Discrete-time Battery Model for System-Level Low-Power Design.
[9] L. Benini, G. Castelli, A. Macii, E. Macii, M. Poncino, R. Scarsi. A Discrete-time Battery Model for High-Level Low-Power Estimation. IEEE 2000 Design Automation and Test in Europe, 35-39, Paris France, March 2000.
[10] L. Benini, G. Castelli, A. Macii, E. Macii, M. Poncino, R. Scarsi. Battery-Driven Dynamic Power Management of Portable Systems. 13th International Symposium on System Synthesis, Madrid, Spain, Sept. 20-22, 2000.
[11] C.F. Chiasserini, R.R. Rao. A Model for Battery Pulsed Discharge with Recovery Effect. IEEE1999.
[12] C.F. Chiasserini, R.R. Rao. Energy Efficient Battery Management. IEEE INFOCOM 2000.
[13] C.F. Chiasserini, R.R. Rao. Battery Life Estimation of Mobile Embedded Systems.
[14] C.F. Chiasserini, R.R. Rao. Pulsed Battery Discharge in Communication Devices, Proceedings of MobiCom 99, August 1999, 88-95.
[15] T. Šimunić, L.Benini, G. De Micheli. Energy-Efficient Design of Battery-Powered Embedded Systems. ISLPED '99, International Symposium on Low Power Electronics and Design Proceedings, 1999.
[16] CoWare N2C User Manual – V2.2.

New “Learning-based” Models of Sub-threshold Bandgap

Salvatore Zappalà

STMicroelectronics S.r.l.
Stradale Primosole 50, 95100 Catania – Italy

Introduction

The goal of this thesis is to realize a mathematical model of sub-threshold bandgap. It became necessary to realise this model due to a bandgap circuit, produced by ST-microelectronic that worked badly in sub-threshold zone.

After having checked and having examined the circuit, it was deduced that the problem was in the model of the MOS used, the MM9; probably it is no longer suitable and too close to its structural limits. So as not to change the product entirely the decision was taken to realise a numerical model that would be able to solve the sub-threshold problems. Firstly, a polynomial model was realized, obtained by working on equations, but good results were not obtained. Consequently a *neural model* was realized, starting with experimental data which was necessary to train the nets.

A huge amount of MOS measures was carried out at ST-microelectronic laboratory, situated in M5, by means of *cell bench* and *parameter analyser.*
When this phase ended, the experimental data was inserted into an appropriate program that had been realized for this specific problem in a MATLAB® area.
The learning rule used was the Back propagation, only one hidden layer that was compounded of 14 neurons was foreseen in the structure.

Firstly the neural nets were trained by using only one MOS data, surveyed at 27°C, in this way, good results and an excellent model was achieved. Successively the averages of a lot of MOS measurements were inserted, at 27°C, 50°C, 75°C and 100°C. Excellent results which were achieved with only one MOS, were confirmed. Then the temperature averages were inserted in the program by using three inputs and one output (V_S, V_G, T) rather than two inputs and one output. After checking that the net was able to run those data excellently and with a very low percentage error, only 10^{-8}, the last phase; the testing phase; was reached.

At first random values were produced to make random net inputs. Some of these inputs were chosen for the learning phase. The remaining data were used to check the outputs and to verify that they were coherent with experimental data. Also in this case the accuracy rate of the neural model has been very high while the error factor has been very low (10^{-8}).

The results obtained are even more significant because, from the model found, a mathematical equation has been extracted, that can be used for the simulation on work-station.

Neural Nets

Artificial neural net models, have been studied for many years to achieve human-like performance in the fields of speech and image recognition.

Neural networks are composed of simple elements operating in parallel. These elements are inspired by biological nervous systems. As in nature, the network function is determined largely by the connections between elements. We can train a neural network to perform a particular function by adjusting the values of the connections (weights) between elements.

Commonly neural networks are adjusted, or trained, so that a particular input leads to a specific target output. Such a situation is shown below. There, the network is adjusted, based on a comparison of the output and the target, until the network output matches the target. Typically many such input/target pairs are used, in this supervised learning, to train a network.

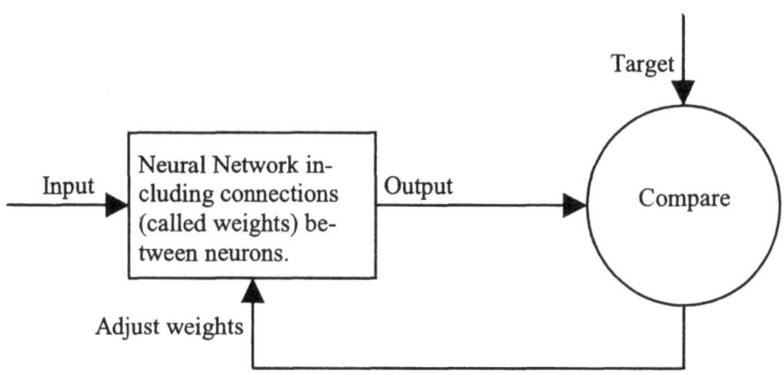

Figure 1. Neural network flow chart

Neural networks have been trained to perform complex functions in various fields of application including pattern recognition, identification, classification, speech, vision and control systems. Today neural networks can be trained to solve problems that are difficult for conventional computers or human beings.

Throughout the development emphasis is placed on neural network paradigms that build up to or are themselves used in engineering, financial and other practical applications. Supervised training methods are commonly used, but other networks can be obtained from unsupervised training techniques or from direct design-methods. Unsupervised networks can be used, for instance, to identify groups of data. Certain kinds of linear networks and Hopfield networks are designed directly.

Neural Nets: History

Neural nets, go by many names such as connectionist and models, parallel distributed processing models. Whatever the name, all these models attempt to achieve good performance via dense interconnection of simple computational elements. Work regarding neural net models has a long history:

1943 McCulloch and Pitts developed a first detailed mathematical model.

1949 Hebbian learning rule.

1954 The first neural computers were invented.

1960 Adaline (a new machine). Widrow–Hoff learning rule (1962).

1969 Marvin Minsky and Seymon Popert.

1982 John Hopfield with his dinamic associative neural net.

1986 Back Propagation.

Today greater interest is due to the development of new net topologies and algorithms for VLSI implementation techniques and some interesting demonstrations as well as by a growing fascination how the human brain functions.

Some Artificial Neuron Models

McCulloch and Pitts' Model (1943)

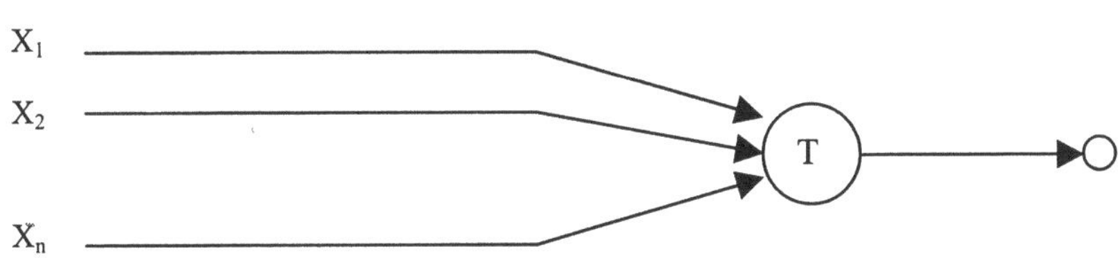

Figure 2. McCulloch and Pitts' model

x_i inputs $\forall$ i=1,2,……,n. Its values can be 0 or 1.

$$O^{K+1} = \begin{cases} 1 \text{ se } & \sum_{n}^{n} w_i^k x_i^k \geq T \\ 0 \text{ se } & \sum_{1}^{n} w_i^k x_i^k \langle\, T \end{cases}$$

where w_i stands for weight of connections and T is the neuron threshold.

Feedforward Model

A single-layer network of S logsig neurons having R inputs is shown below in full detail on the left and with a layer diagram on the right.

Feedforward networks often have one or more hidden layers of sigmoid neurons followed by an output layer of linear neurons. Multiple layers of neurons with non-linear transfer functions allow the network to learn non-linear and linear relationships between input and output vectors. The linear output layer lets the network produce values outside the range –1 to +1.

On the other hand, if it is desirable to constrain the outputs of a network (such as between 0 and 1) then the output layer should use a sigmoid transfer function (such as logsig).

For multiple-layer networks we use the number of the layers to determine the superscript on the weight matrices. The appropriate notation is used in the two-layer tansig/purelin network shown next.

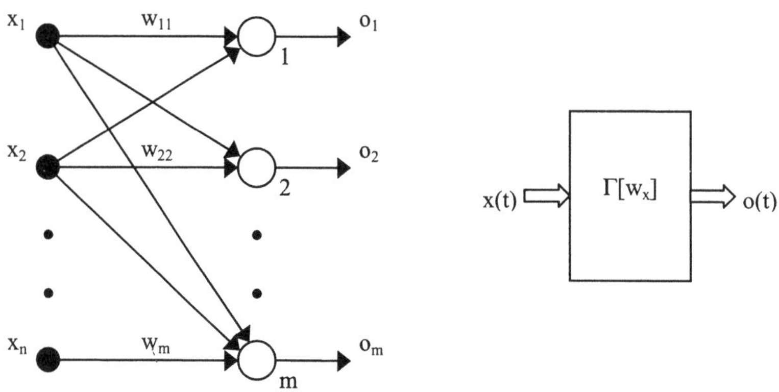

Figure 3. Feedforward model

$o = [o_1, o_2, \ldots\ldots, o_n]^t$ is the output pattern.
$x = [o_1, o_2, \ldots\ldots, o_n]^t$ is the input pattern.

The activation value is:

$$net_i = \sum_{1}^{n} w_{ij} x_j \quad \forall \; i=1,2,\ldots\ldots,m.$$

$o_i = f(w_i^t x)$ $\forall$ I=1,2,......,n and f is a non-linear function.

Retrospective Nets (Feedback)

These nets are made connecting the output with the input:

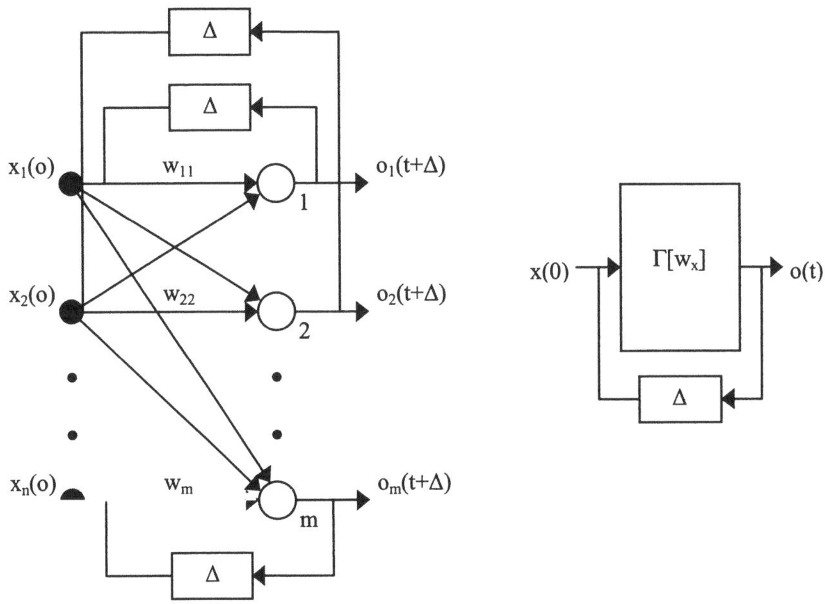

Figure 4. Feedback flow chart

The inputs to a neuron include its bias and the sum of its weighted inputs (using the inner product). The output of a neuron depends on the neuron's inputs and on its transfer function. There are many useful transfer functions. A single neuron cannot do very much. However, several neurons can be combined into a layer or multiple layers that have great power. Hopefully this toolbox makes it easy to create and understand such large networks.

The architecture of a network consists of a description of how many layers a network has, the number of neurons in each layer, each layer's transfer function, and how the layers are connected to each other. The best architecture to use depends on the type of problem to be represented by the network.

A network effects a computation by mapping input values to output values. The particular mapping problem to be performed fixes the number of inputs as well as the number of outputs for the network.

Apart from the number of neurons in a network's output layer, the number of neurons in each layer is up to the designer. Except for purely linear networks, the more neurons in a hidden layer the more powerful the network. If a linear mapping needs to be represented linear neurons should be used.

However, linear networks cannot perform any non-linear computation. Use of a non-linear transfer function makes a network capable of storing non-linear relationships between input and output.

A very simple problem may be represented by a single layer of neurons. However, single layer networks cannot solve certain problems. Multiple feed-forward layers give a network greater freedom. For example, any reasonable function can be represented with a two layer network: a sigmoid layer feeding a linear output layer.

Networks with biases can represent relationships between inputs and outputs more easily than networks without biases. (For example, a neuron without a bias will always have a net input to the transfer function of zero when all of its inputs are zero. However, a neuron with a bias can learn to have any net transfer function input under the same conditions by learning an appropriate value for the bias.)

Feed-forward networks cannot perform temporal computation. More complex networks with internal feedback paths are required for temporal behaviour. If several input vectors are to be presented to a network, they may be presented sequentially or concurrently. Batching of concurrent inputs is computationally more efficient and may be what is desired in any case.

Learning Rules

We will define a learning rule as a procedure for modifying the weights and biases of a network. (This procedure may also be referred to as a training algorithm.) The learning rule is applied to train the network to perform some particular task. Learning rules in this toolbox fall into two broad categories: *supervised learning* and *unsupervised learning*.

In *supervised learning*, the learning rule is provided with a set of examples (the training set) of proper network behaviour:

$$\{p_1\text{-}t_1\}\ ,\ \{p_2\text{-}t_2\}\ ,\ \ldots\ldots.,\ \{p_Q\text{-}t_Q\}$$

where p_i is an input to the network, and t_j is the corresponding correct (target) output. As the inputs are applied to the network, the network outputs are compared to the targets. The learning rule is then used to adjust the weights and biases of the network in order to move the network outputs closer to the targets. The perceptron learning rule falls into this supervised learning category.

In *unsupervised learning*, the weights and biases are modified in response to network inputs only. There are no target outputs available. Most of these algorithms perform clustering operations. They categorize the input patterns into a finite number of classes. This is especially useful in such applications as vector quantization.

As noted, the Perceptron discussed in this chapter is trained with supervised learning. It is to be hoped that a network that produces the right output for a particular input will be obtained.

Perceptron Learning Rule. (Single Layer Perceptron)

The simple layer perceptron can be used with both continuous valued and binary inputs. Connection weight and the threshold in a perceptron can be fixed or adapted using a number of different algorithms.The original perceptron convergence procedure for adjusting weights was developed by Roseblatt. The error is:

$$r = d_i - o_I \qquad \begin{cases} d_i = \textit{wanted output} \\ o_i = \operatorname{sgn}(w_i^j x) \end{cases}$$

First connection weights and the threshold value are initialised to small random non-zero values. Then a new input with n continuous valued elements is applied to the input and to the output is computed.

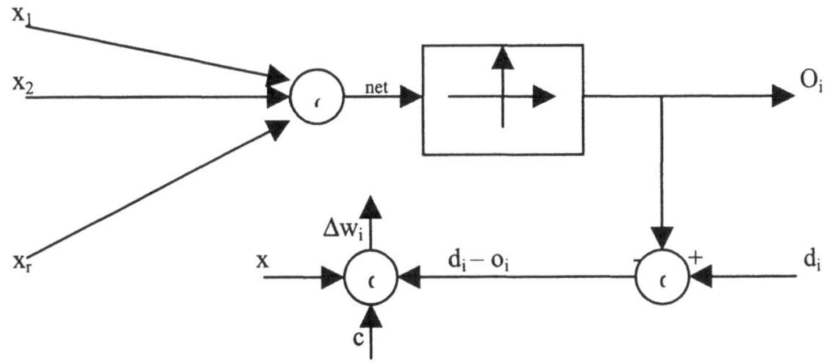

Figure 5. Perceptron learning rule flow chart

Connection weights are adapted only when an error occurs.

$$\Delta w_i = c\,[\,d_i - \text{sgn}\,(\,w_i^t\ x\,)]\ x$$
$$\Delta w_{ij} = c\,[\,d_i - \text{sgn}\,(\,w_i^t\ x\,)]\ x_j \qquad \forall\ j = 1, 2, \ldots, n.$$

The actual output is:

$$y(t) = f_n\left(\sum_{i=1}^{n} w_i(t)x_i(t) - \theta\right)$$

Multi-layer Perceptron. (BackPropagation)

Backpropagation was created by generalizing the Widrow-Hoff learning rule to multiple-layer networks and non-linear differentiable transfer functions. Input vectors and the corresponding output vectors are used to train a network until it can approximate a function, associate input vectors with specific output vectors, or classify input vectors in an appropriate way as defined by us. Networks with biases, a sigmoid layer, and a linear output layer are capable of approximating any function with a finite number of discontinuities. Standard backpropagation is a gradient descent algorithm, as is the Widrow-Hoff learning rule.

The term backpropagation refers to the manner in which the gradient is computed for non-linear multi-layer networks. There are a number of variations on the basic algorithm which are based on other standard optimisation techniques, such as conjugate gradient and Newton methods. Properly-trained backpropagation networks tend to give reasonable answers when presented with inputs that they have never seen. Typically, a new input will lead to an output similar to the correct output for input vectors used in training that are similar to the new input being presented. This generalization property makes it possible to train a network on a representative set of input target pairs and get good results without training the network on all possible input/output pairs.

The sum of the weighted inputs and the bias forms the input to the transfer function f. Neurons may use any differentiable transfer function f to generate their output. The most used transfer function is the sigmoid function:

$$f(net) = \frac{1}{1 + \exp(-\sum_j w_{ij} o_j + \vartheta_i)}$$

In this section we want to present the architecture of the network which is most commonly used with the backpropagation algorithm the multi-layer feed forward network. The routines in the Neural Network Toolbox can be used to train more general networks.

Neuron Model (TANSIG, LOGSIG, PURELIN)

An elementary neuron with R inputs is shown below. Each input is weighted with an appropriate w. The sum of the weighted inputs and the bias forms the input to the transfer function f. Neurons may use any differentiable transfer function f to generate their output.

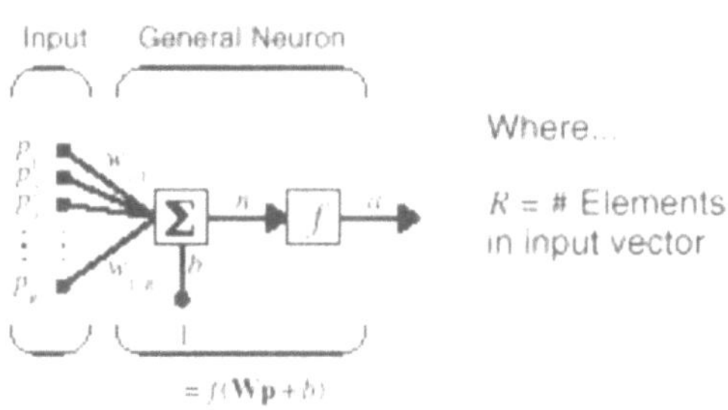

Figure 6. Neuron model flow chart

Multi-layer networks often use the log-sigmoid transfer function logsig.

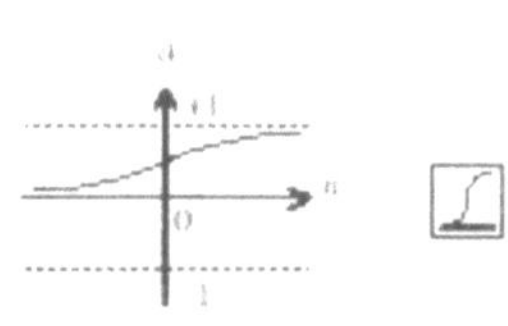

Figure 7. Log-sigmoid transfer function.

The function logsig generates outputs between 0 and 1 as the neuron's net input goes from negative to positive infinity. Alternatively, multi-layer networks may use the tan-sigmoid transfer function tansig.

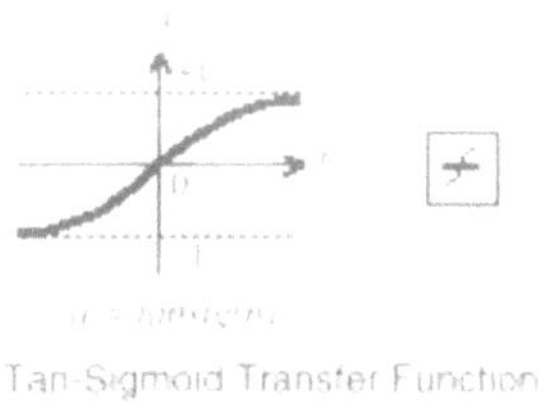

Figure 8. Tan-sigmoid transfer function

Occasionally, the linear transfer function purelin is used in backpropagation networks.

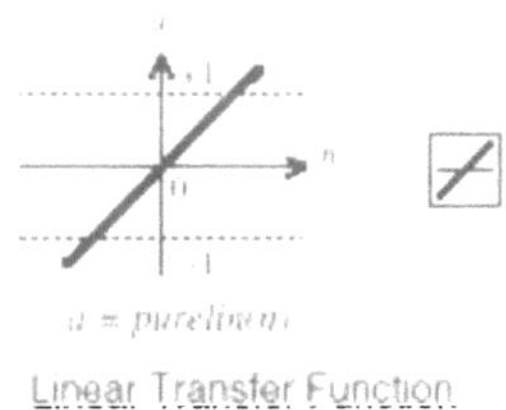

Figure 9. Linear transfer function

If the last layer of a multi-layer network has sigmoid neurons, then the outputs of the network are limited to a small range. If linear output neurons are used the network outputs can take on any value.

The Back-propagation Training Algorithm

The Back-propagation training algorithm is an iterative gradient algorithm designed to minimize the mean square error between the actual output of a multi-layer feed-forward perceptron and the desired output.
It requires continuous differentiable non-linear ties.

Step 1. Initialise weights and offsets.
Set all weights and node offsets to small random values.

Step 2. Present input and desired output.
Present a continuous input vector x_0, x_1, …,x_{n-1} and specify the desired outputs d_0,d_1, …,d_{m-1}. If the net is used as a classifier then all desired outputs are typically set to zero except for the one corresponding to the class the input is from. That desired output is 1. The input could be new on each trial or sample from a training set could be presented cyclically until weights stabilize.

Step 3. Calculate actual outputs.
Use the sigmoid non-linearity to calculate outputs y_0, y_1, …y_{M-1}.

Step 4. Adapt weights.
Use a recursive algorithm starting at the output nodes and working back to the first hidden layer. Adjust weights by:

$$W_{ij}(t+1)=w_{ij}(t)+\eta\delta_j x_i^{'}$$

In this equation $w_{ij}(t)$ is the weight from hidden node i or from an input to node j. the error term for node j is:

$$\delta_j = y_i(1-y_j)(d_i-y_j).$$

if node j is an internal hidden node, then

$$\delta_j = x_i^{'}(1-x_j^{'})\sum_K \delta_k w_{jk}.$$

Step 5. Repeat by going to step 2.

CMOS Bandgap Reference.

The fast progress in MOS VLSI technologies allows the implementation of increasingly sophisticated systems on a single chip. Although these systems are mainly based on digital circuits, they usually need some analogue circuitry to perform functions difficult or impossible to realize by digital means. For economic reasons, these analogue circuits must be realized without tighter control or additional steps in the process.

During the last few years, a series of new solutions were proposed to meet these requirements, mainly in the areas of A/D conversion and filtering. Recently, some voltage reference that uses the difference between gate-source voltages in an enhancement depletion has been described.

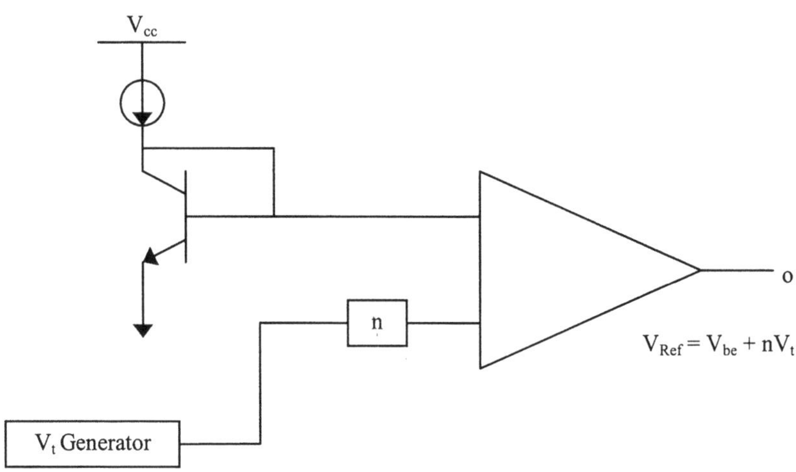

Figure 10. Bandgap voltage reference

Bandgap voltage references are widely used in bipolar technology where they reach very high standards of accuracy and stability. Their principle relies on the fact that the voltage V_j across a p-n junction that is forward biased by a constant current, increases fairly linearly with decreasing temperature and tends toward the bandgap value V_{G0} when the absolute temperature is extrapolated to 0.

If a voltage that is proportional to absolute temperature (PTAT) is added to V_j to compensate exactly the difference between V_j and V_{G0}, one obtains a total voltage that is independent of the temperature.

The realization of bandgap reference in MOS technology has long been hindered by the difficulty of realizing a PTAT voltage. Solutions to this problem may be found by considering the behaviour of a MOS transistor in weak inversion. For a negligible density N_{ss} of fast surface states, the drain current may be expressed as:

$$I_D = SI_{D0}e^{\frac{V_G}{nU_T}}\left(e^{\frac{-V_S}{U_T}} - e^{\frac{-V_D}{U_T}}\right) \tag{1}$$

where S is the effective geometrical shape factor of the transistor; V_G, V_S, and V_D are the gate, source and drain potentials with respect to the substrate, and $U_T = kT/q$. the slope factor n is fairly controllable, whereas the characteristic current I_{D0} is very sensitive to process and temperature.

It is well known that when the gate – source voltage of a MOS transistor is reduced below the threshold voltage defined by the usual strong inversion characteristics, the channel current decreases approximately exponentially. This sub-threshold region of the characteristics, inside which the device operates with a weakly inverted channel, has been studied by many authors.

Barron has developed the pertinent theory and has obtained a solution in close form by introducing some approximations.

Swanson and Meindl have elaborated a similar expression in order to describe the behaviour of MOS inverters at a low voltage.

Troutman and Chakravarti have derived a model valid for any substrate bias and extendable to short channel lengths; they have shown that channel current, in weak inversion, flows by diffusion.

Van Overstraeten has demonstrated that this diffusion current is a function of the inversion charge at the source and drain, and has pointed out the influence of surface potential fluctuations on the I_D-V_G characteristics.

Recently, Barker has proposed to use weak inversion for small signal amplification and has derived an appropriate model.

Principle of Reference

Let us make the following assumptions:

- the channel is sufficiently long so that the gradual channel approximation can be used and channel length modulation effects are negligible;
- generation current in the drain, channel, and source depletion regions are negligible; source and drain currents are then equal;

- the density of fast surface states and the fluctuations of surface potential are negligible.

The basic derivation of Barron may then be used and can be easily extended to the case of non-zero source – substrate voltage. His approximate expression for the weak inversion current may be rewritten, for an n-channel transistor, as:

$$I_D = S\mu \ U_T^2 \left(\frac{1}{2} q\varepsilon_S n_i \right)^{1/2} e^{-\left(\frac{3\phi}{2U_T}\right)} \frac{e^{\Psi_S / U_T}}{(\Psi_S - U_T)^{1/2}} (e^{\frac{-V_S}{U_T}} - e^{\frac{-V_D}{U_T}}) \tag{2}$$

where:

S = geometrical shape factor of the transistor;
μ = mobility of carriers in the channel;
$U_T = kT/q$;
ϵ_S = permittivity of S_i.
$\Phi = U_T \ln(N_B/n_i)$ bulk Fermi potential,
N_B = constant bulk impurity concentration;
n_i = Intrinsic carrier concentration;
ψ_S = surface potential, constant along the channel in weak inversion;
V_S = source – substrate voltage;
V_D = drain – substrate voltage;
V_G = gate – substrate voltage;
I_D = drain current.

It is valid for:

$$4U_T + \Phi + V_S < \psi_S < 2\Phi + V_S.$$

The surface depletion capacitance C_d can be expressed as:

$$C_d = \left(\frac{\frac{1}{2} \varepsilon_S q n_i}{\Psi_S - U_T} \right)^{1/2} e^{\Phi / 2U_T}$$

Then the expression (2) yields:

$$I_D = S\mu \ U_T^2 e^{-\left(\frac{2\phi}{2U_T}\right)} e^{\frac{\psi_S}{U_T}} (e^{\frac{-V_S}{U_T}} - e^{\frac{-V_D}{U_T}})$$

Owing to the very slow variation of C_d with ψ_S, I_D is essentially dependent exponentially on ψ_S/U_T.

Variations of the gate – substrate voltage V_G are shared between the oxide capacitance per unit area C_{0x} and the semiconductor total surface capacitance per unit area C_s. if C_G is the gate capacitance per unit area, we have:

$$\frac{\partial \Psi_S}{\partial V_G} = \frac{C_{0x}}{C_{ox} - C_S} = 1 - \frac{C_G}{C_{0x}}$$

C_G is fairly constant for: $4U_T + \Phi + V_S < \psi_S < 2\Phi + V_S - 2U_T$.

Inside this range ψ_S is linearly dependent on V_G, then I obtain the equation (1):

$$I_D = SI_{D0} e^{\frac{V_G}{nU_T}} (e^{\frac{-V_S}{U_T}} - e^{\frac{-V_D}{U_T}})$$

where I_{D0} is a characteristic current and n a slope factor. As a negligible density of fast surface states it is assumed, C_S is equal to C_d within the range of interest and yields the slope factor:

$$n = 1 + \frac{C_d}{C_{0x}} \tag{3}$$

We can evaluate n at the minimum of the C_G-V_G curve, then $\psi_S \approx \Phi + V_S - 4U_T$, then:

$$n = \frac{1}{1 - \left({C_G}/{C_{0x}} \right)\min} = 1 + \frac{1}{C_{0x}} \left(\frac{qN_B \varepsilon_S}{2(2\Phi - 5U_T + V_S} \right)^{1/2}.$$

Thanks to the slow variation of C_d with ψ_S, the slope factor n may be considered as a constant for transistors biased by similar values of V_S.

The same is true for the characteristic current I_{D0} for transistors on the same chip. I_{d0} is very poorly controlled from batch to batch and its absolute value is of no use to the circuit designer; the designs must be based on drain current ratios, as is common practice for bipolar circuits. On the contrary, n is a reproducible parameter available for circuit design.

As shown, n takes different values for transistors biased by widely different values of V_S. the same is true for I_{D0}, so that the technique of drain current ratios can only be applied to transistors with differential values of V_S not exceeding a few U_T. This is not a real limitation to design flexibility.

Expression (1), is seen to be symmetrical in V_D and V_S, as can be expected from the symmetrical structure of the device. It is applicable to p-channel transistors as well as by changing the signs of V_G, V_S and V_D.

The transconductance in weak inversion is:

$$g_m = \frac{\partial I_D}{\partial V_G} = \frac{I_D}{nU_T}$$

For V_D-V_S<<U_T, a source transconductance can be defined as:

$$g_{mS} = \frac{\partial I_D}{\partial V_S} = \frac{I_D}{U_T}$$

The upper limit for (1) is obtained by introducing the upper limit of ψ_S and replacing C_D by its value taken from (3); then for V_D-V_S>3U_T (saturated drain current), we have:

$$I_D \leq \frac{n-1}{e^2} S\mu C_{0x} U_T^2 \text{ (Limit of validity in weak inversion region)}$$

Sub-threshold Gradient Slope

It has been found from drain that:

$$n = e^{\alpha_1 I + \beta_1 I^2 + \gamma_1 I^3}$$

Now it is necessary to prove that n is valid also when the MOS is seen from source (V_{GS}=0). Now, V_D=1.5V, V_S=-0.1 – 0.1 V, V_{GS}=0.3 – 0.7 V. n, determines the slope of the sub-threshold curve.

It is then important to prove if n is suitable or if it is necessary to find a new dependence: n(I, V_{SB}).

Before it was

$$I_D = SI_{D0} e^{\frac{V_G}{nU_T}} (e^{\frac{-V_S}{U_T}} - e^{\frac{-V_D}{U_T}}) \tag{1}$$

and was reduced to:

$$I_D = SI_{D0} e^{\frac{V_{GB}}{nU_T}}$$

because $(e^{\frac{-V_S}{U_T}} - e^{\frac{-V_D}{U_T}}) \approx 1$.

Now, as the value of V_{GS} is known, it is considered:

$$I_D = SI_{D0} e^{\frac{V_{GS}+V_{SB}}{nU_T}} (e^{\frac{-V_{SB}}{U_T}} - e^{\frac{-V_{DB}}{U_T}}).$$

The drain is fixed 1.5V and (1) is equivalent to:

$$I_D = SI_{D0} e^{\frac{V_{GS}-V_{SB}}{\eta U_T}} (1 - e^{\frac{-V_{DB}}{U_T}}) \tag{2}$$

then sub-threshold gradient is calculated:

$$\frac{1}{S}=\frac{d\log_{10} I_D}{dV_{GS}}$$

Then (2) is integrated:

$$\ln I_D = \ln I_0 + \frac{V_{GS}-V_{th}}{\eta V_T},$$

$$\log_{10} I = \frac{\ln I}{\ln 10} = \frac{\ln I_0}{\ln 10} + \frac{1}{\ln 10}\frac{V_{GS}-V_{th}}{\eta V_T},$$

$$\frac{d\log_{10} I_D}{dV_{GS}} = \frac{1}{\ln 10}\frac{1}{\eta k}.$$

It was noticed that μ expresses the same thing as n.

$S = \ln 10\, \eta\, V_T$	<u>Sub-threshold slope</u>.

It was noticed that fixing the gate and moving the source, everything works as before and n depends only on the temperature.

Laboratory Measurement: Measures

In this chapter a map about the measures carried out and the formality followed will be introduced.

The measures were carried out at STMicroelectronics laboratories situated in M5, by means of *cell bench* and *parameter analyser*. All the temperatures were checked and the *cell bench* was shielded from external agents.

To reduce the error margin to a minimum, the microtips were very carefully placed. The wafer on which these measures were effected is composed of about 64 MOS in technology F6Y with W=5μ and L=2.5μ.

Measurement map at 27°C

(V_S=-0.1:0.001:0.1V; V_G=0.3V:0.1V:0.7V; Hold=100ms.)
(V_D=1V fixed.)

Carried out measure   Failed measure   Not effected measure

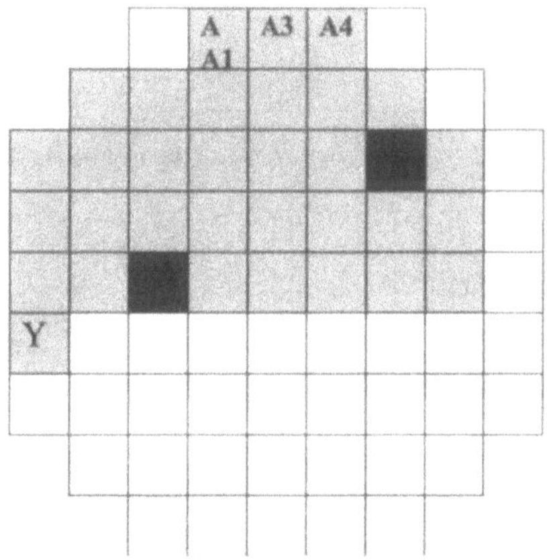

Measurement map at various temperatures

(V_S=0 : 2mV : 100mV; V_G=0.3V : 50m V : 0.7V)
(V_D=1.5V fixed, Hold=100ms.)

T=27°C

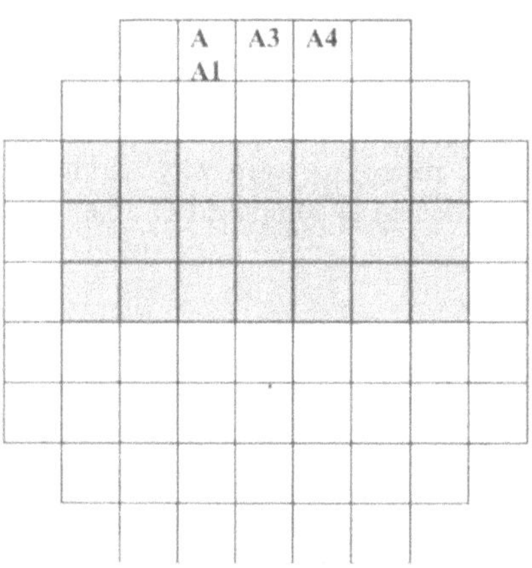

T=50°C

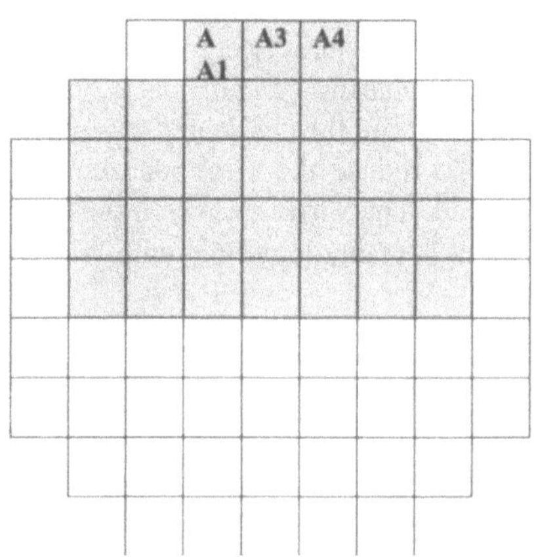

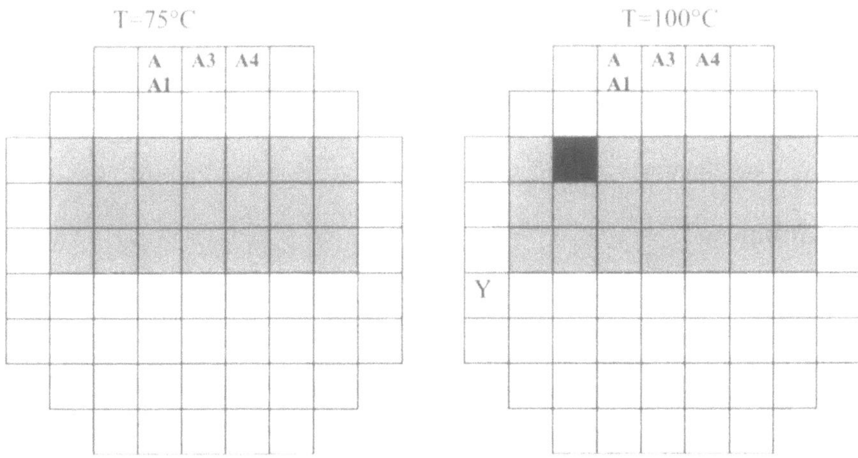

Learning-based Model

A program was realized in MATLAB® to implement a neural algorithm.
he net, that was used, is composed of:

- three inputs (V_S, V_G, T).
  The temperature averages at 27°C, 50°C, 75°C and 100°C, were inserted in the program;
- 14 hidden layers.
  This "hidden layer" layers number was chosen after various trials, because it was the one which gave the best results and which obtained optimum elaboration times-conditions;
- one output: I_D.
  The output is composed of the only value of I_D, as it is sufficient to characterize the sub-threshold slope.

Program Details

As we said the learning rule used in this program is the back propagation. Back propagation was created by generalizing the Widrow-Hoff learning rule to multiple-layer networks and non-linear differentiable transfer functions. Input vectors and the corresponding output vectors are used to train a network until it can approximate a function, associate input vectors with specific output vectors, or classify input vectors in an appropriate way as defined by us.

Networks with biases, a sigmoid layer, and a linear output layer are capable of approximating any function with a finite number of discontinuities.

Standard back propagation is a gradient descent algorithm, as is the Widrow-Hoff learning rule.

There are a number of variations on the basic algorithm which are based on other standard optimisation techniques, such as conjugate gradient and Newton methods. The Neural Network Toolbox implements a number of these variations.

Before training a feed forward network, the weights and biases must be initialised. The initial weights and biases are created with the command init. This function takes a network object as input and returns a network object with all weights and biases initialised. Here is how a network is initialised:

net = init(net).

Training

Once the network weights and biases have been initialised, the network is ready for training. The network can be trained for function approximation (non-linear regression), pattern association, or pattern classification.

The training process requires a set of examples of proper network behaviour network inputs p and target outputs t. During training the weights and biases of the network are iteratively adjusted.

Now we will describe several different training algorithms, we tested, for feed forward networks.

Batch Training (TRAIN)
The batch training, is invoked using the function train. In batch mode the weights and biases of the network are updated only after the entire training set has been applied to the network.

The gradients calculated at each training example are added together to determine the change in the weights and biases. The principal function to improve it is:

Batch Gradient Descent (TRAINGD).
It implements the batching form of the standard steepest descent training function. The weights and biases are updated in the direction of the negative gradient of the performance function.

Batch Gradient Descent With Momentum (TRAINGDM).
The batch form of gradient descent with momentum is invoked using the training function traingdm.

These two algorithms gave an equivalent error, about 10^{-5}, insufficient for the required precision. So it was necessary to look for different algorithms.

Faster Training

The previous section presented two back propagation training algorithms: gradient descent and gradient descent with momentum. These two methods are often too slow for practical problems.

Now we will discuss several high-performance algorithms which can converge from ten to one hundred times faster than the algorithms discussed previously. All of the algorithms in this section operate in the batch mode and are invoked using train. These faster algorithms fall into two main categories:

- the first category uses heuristic techniques, which were developed by analising the performance of the standard steepest descent algorithm. We

tested two heuristic techniques: variable learning rate back propagation, *traingda,* and resilient back propagation *trainrp*;
the second category of fast algorithms uses standard numerical optimisation techniques. In this phase we tested three types of numerical optimisation techniques for neural network training: conjugate gradient (*traincgf, traincgp, traincgb, trainscg*), quasi-Newton (*trainbfg, trainoss*), and Levenberg-Marquardt (*trainlm*).

Variable Learning Rate (TRAINGDA, TRAINGDX)
With standard steepest descent, the learning rate is held constant throughout training. The performance of the algorithm is very sensitive to the proper setting of the learning rate. If the learning rate is set too high, the algorithm may oscillate and become unstable. If the learning rate is too low, the algorithm will take too long to converge. It is not practical to determine the optimal setting for the learning rate before training, and, in fact, the optimal learning rate changes during the training process, as the algorithm moves across the performance surface.

These two algorithms gave a variable error, about $10^{-5} \div 10^{-6}$, insufficient for the required precision. Better results were obtained using trainrp function.

Resilient Back propagation (TRAINRP)
Multi-layer networks typically use sigmoid transfer functions in the hidden layers. These functions are often called squashing functions, since they compress an infinite input range into a finite output range. Sigmoid functions are characterized by the fact that their slope must approach zero as the input gets large. This causes a problem when using steepest descent to train a multi-layer network with sigmoid functions, since the gradient can have a very high magnitude, and therefore cause small changes in the weights and biases, even though the weights and biases are far from their optimal values.

The purpose of the resilient back propagation (Rprop) training algorithm is to eliminate these harmful effects of the magnitudes of the partial derivatives. In this case the error rate was very low, about $10^{-6} \div 10^{-7}$.

Conjugate Gradient Algorithms
The basic back propagation algorithm adjusts the weights in the steepest descent direction (negative of the gradient). This is the direction in which the performance function is decreasing most rapidly. It turns out that, although the function decreases most rapidly along the negative of the gradient, this does not necessarily produce the fastest convergence. In the conjugate gradient algorithms a search is performed along conjugate directions, which produces generally faster convergence than steepest descent directions. We checked four different variations of conjugate gradient algorithms; *traincgf, traincgp, traincgb, trainscg*. These functions were unsuitable to solve the problem and they gave a high error rate, about 10^{-5}.

The results obtained using quasi-Newton algorithms; BFGS algorithm (TRAINBFG) and one-step secant algorithm (TRAINOSS) were not adequate.

Levenberg-Marquardt (TRAINLM)

The best results were obtained using Levenberg-Marquardt (*TRAINLM*).Like the quasi-Newton methods, the Levenberg-Marquardt algorithm was designed to approach second-order training speed without having to compute the Hessian matrix. When the performance function has the form of a sum of squares (as is typical in training feed forward networks), then the Hessian matrix can be approximated as:

$$H=\mathbf{J}^T\mathbf{J}$$

and the gradient can be computed as:

$$g=\mathbf{J}^T e$$

where **J** is the Jacobian matrix, which contains first derivatives of the network errors with respect to the weights and biases, and **e** is a vector of network errors. The Jacobian matrix can be computed through a standard backpropagation technique that is much less complex than computing the Hessian matrix. This algorithm appears to be the fastest method for training moderate-sized Feedforward neural networks (up to several hundred weights). It also has a very efficient MATLAB® implementation, since the solution of the matrix equation is a built-in function, so its attributes become even more pronounced in a MATLAB® setting.

The Levenberg-Marquardt algorithm uses this approximation to the Hessian matrix in the following Newton-like update:

$$x_{k+1} = x_k - (J^T J + \mu I)^{-1} J^T e$$

When the scalar m is zero, this is just Newton's method, using the approximate Hessian matrix. When m is large, this becomes gradient descent with a small step size. Newton's method is faster and more accurate near an error minimum, so the aim is to shift towards Newton's method as quickly as possible. Thus, m is decreased after each successful step (reduction in performance function) and is increased only when a tentative step would increase the performance function. In this way, the performance function will always be reduced at each iteration of the algorithm. In this case the error was negligible, about 10^{-8}. Therefore the *trainLM algorithm* was used successfully obtaining excellent results.

In the learning phase not all the experimental data were used but only ¾ of them, chosen at random, the remaining data were used to check the program. Owing to this phase the results were excellent and the curves of the currents at various temperatures fitted perfectly the experimental curves. Notice that the visible holes were due to the missing data that were later used to check the program.

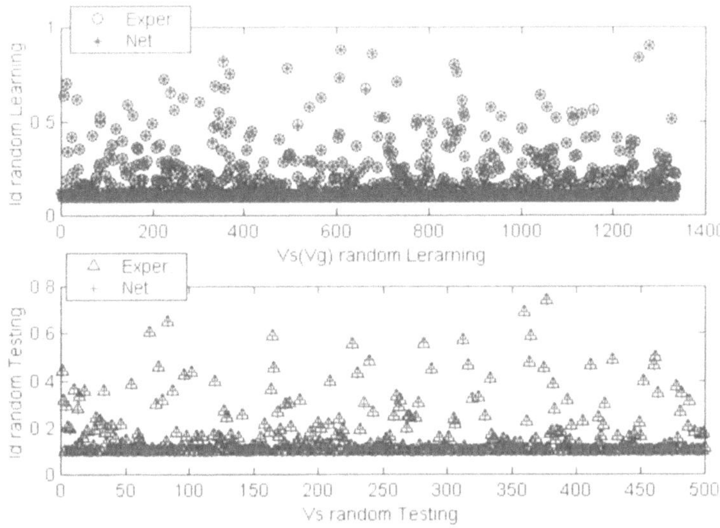

Figure 11. Random learning and testing results

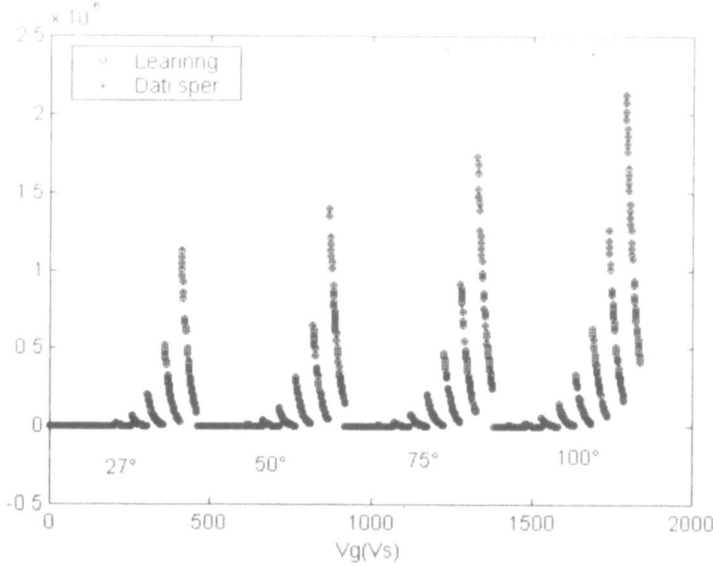

Figure 12. Learning and experimental results

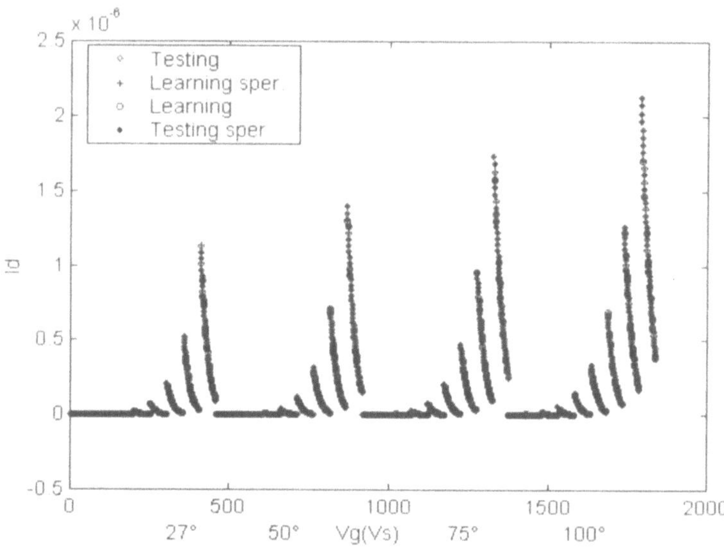

Figure 13. Learning experimental and testing results

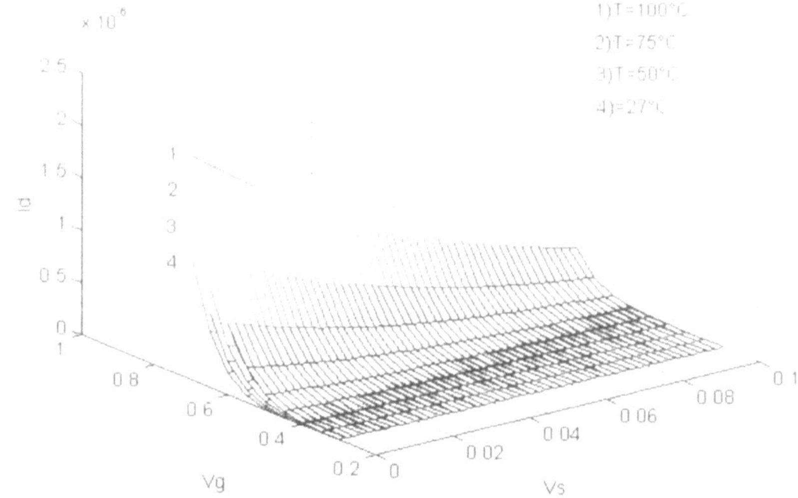

Figure 14. Three dimension learning and testing results

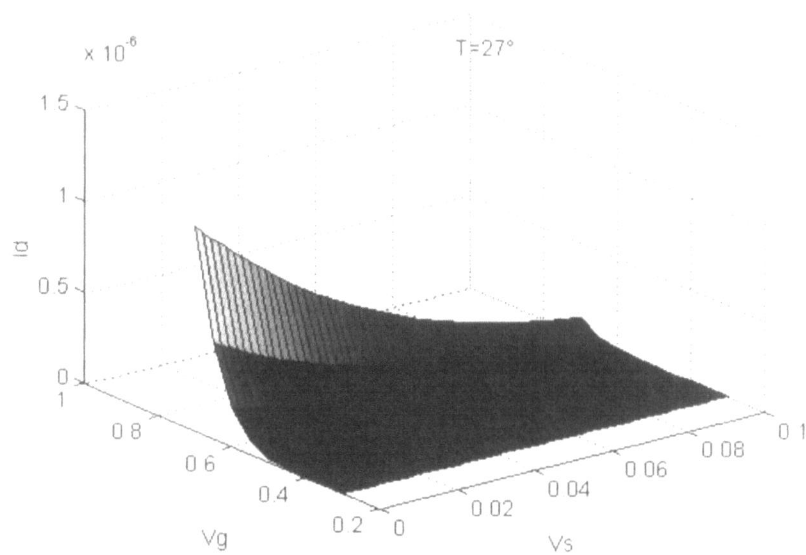

Figure 15. Three-dimensional results at 27°C

Figure 16. Level curves

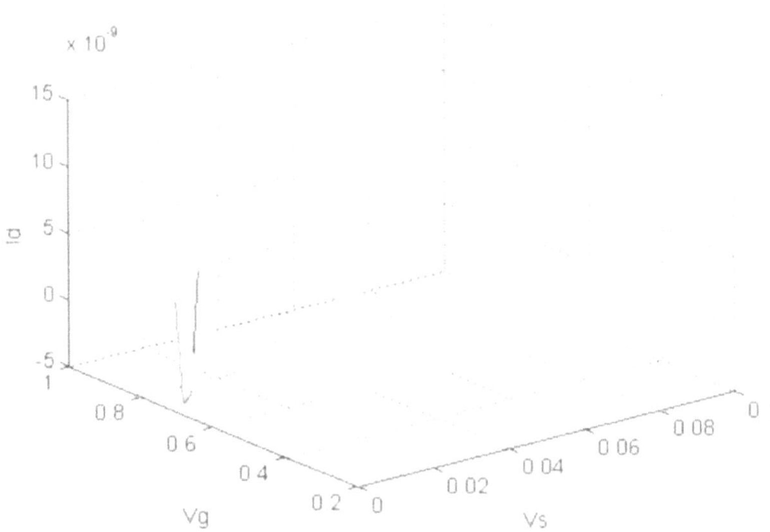

Figure 17. Percentage of error graph

After this checking phase we passed to the settlement of the sub-threshold slope, giving particular attention to the most significant values ($1 \div 100nA$).

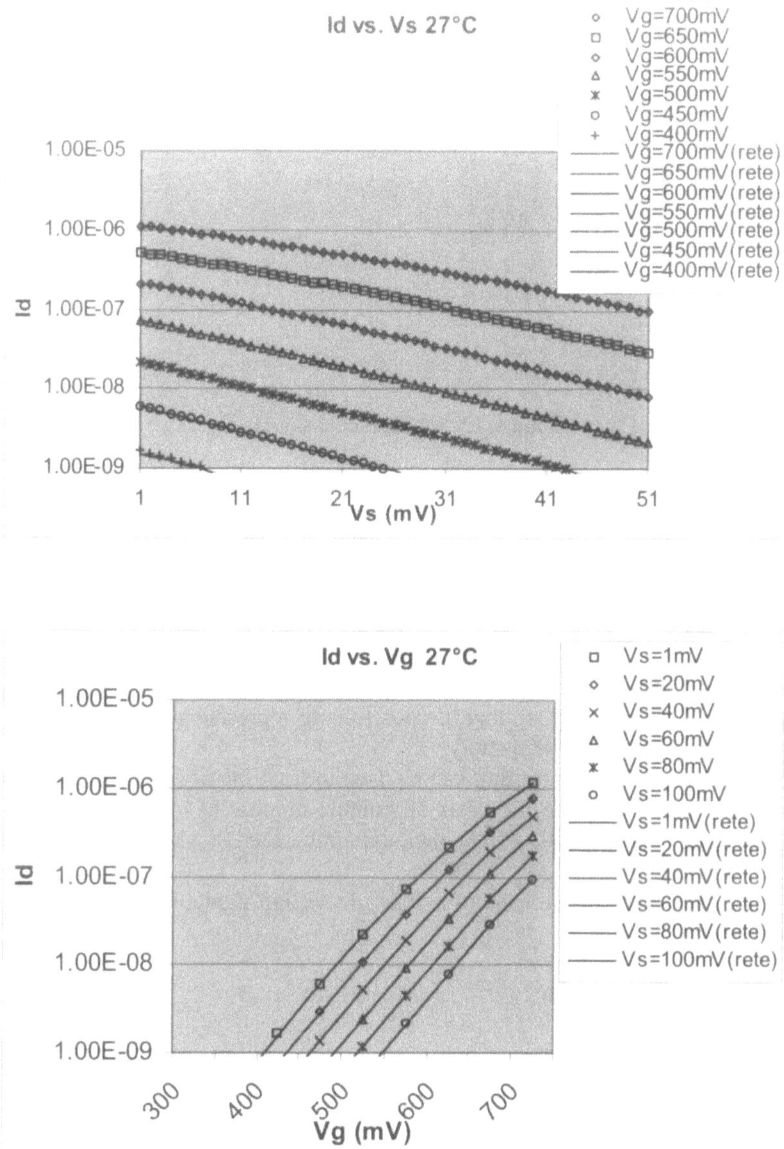

Figure 18. Sub-threshold slopes

At this point it was suitable to pass again to the settlement of the sub-threshold slope, considering: $S = \dfrac{dV_G}{d\log_{10} I_D}$.

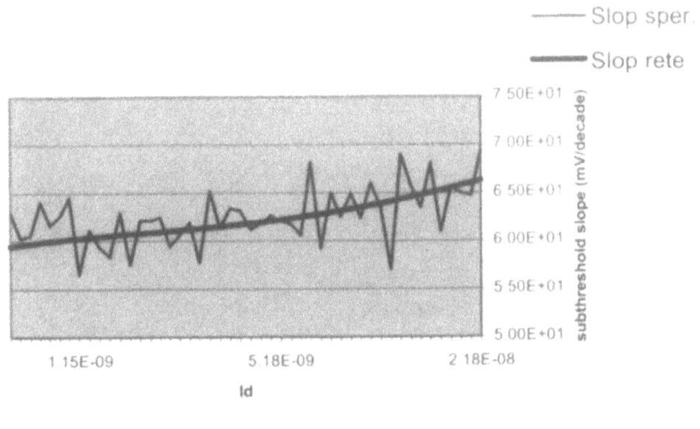

Figure 19. Sub-threshold slopes

The results obtained are excellent, the matching is perfect, or better still the experimental data were clear of errors.

Model Approximation with Multi-layer Perceptron

Multi-layer perceptrons can approximate any non-linear continuous function with an arbitrary degree of accuracy, so that it's possible, to represent any non-linear model by choosing their structure adequately.

The back propagation algorithm can be used to train the neural network to simulate the system dynamics on the basis of suitable number of training patterns, that can be considered as examples of the plant dynamics, then it can be considered as an identification strategy.

A multi-layer perceptron leads to a dynamic system model by using the following expression:

$$N_i(U) = \Gamma[W_i U + B_i]$$

used in a recursive manner;

U is the output of the (i-1)th layer;
W_i is the weight matrix associated to the i-th layer;
B_i is the bias vector corresponding to the i-th layer;
Γ is an appropriate vector valued function that depends on the activation function used. The output is:

$$Y = F(U) = \Gamma(W_n(\Gamma(W_{n-1}(\ldots.\Gamma(W_1U)))))$$

The biases have not been reported for the sake of simplicity.

It is known that a multi-layer perceptron with one hidden layer can solve the same class of problems that can be solved by being used with several hidden layers. Then we could use the following expression:

$$Y = \Gamma(W_2(\Gamma(W_1U+B_1))+B_2).$$

The expression reported above represents the class of models that is obtained when a dynamic system is identified by using multi-layer perceptrons, based on a suitable choice of the input and output structure.

The weights and bias vectors are:

$$W_1 = \begin{bmatrix} -0.1015 & 0.2467 & -3.4224 \\ -0.0010 & -2.3640 & -0.0007 \\ -0.0339 & -0.3669 & -0.1187 \\ -0.0567 & 0.1606 & 0.0371 \\ -0.1762 & 0.5034 & 0.0355 \\ 0.0573 & -0.1635 & -0.0469 \\ -0.0013 & -4.5329 & 0.3520 \\ 0.8706 & 13.6405 & -1.5041 \\ 0.0691 & -0.1723 & -0.6709 \\ 0.2233 & -2.4134 & -0.0776 \\ 0.1772 & -0.5059 & -0.0353 \\ 0.2171 & -0.4929 & -0.0782 \\ -0.0931 & 0.2715 & 0.0299 \\ -0.0666 & 0.1679 & -1.9175 \end{bmatrix}$$

$$W_2=[\,-3.3749\quad 21.2192\quad 0.0001\quad -0.0047\quad -1.2901\quad -0.0972\quad -0.0159$$

$$-0.0022\quad -0.0594\quad -44.6466\quad -1.2521\quad -0.0956\quad 0.3132\quad 3.3749\,]$$

In this case, U is constant and is the inputs vector, containing V_G, V_S and T.

$$B_1 = \begin{bmatrix} 4.9989 \\ 27.3235 \\ 0.8446 \\ -0.2786 \\ -4.7312 \\ 1.6738 \\ 40.1107 \\ -128.1491 \\ 8.8980 \\ 29.7189 \\ 4.7520 \\ 7.0162 \\ -3.0807 \\ 1.3834 \end{bmatrix} \qquad B_2 = 24.0642.$$

and :

$$\Gamma(\cdot) = \begin{bmatrix} f(\cdot) \\ f(\cdot) \\ \ldots\ldots \\ f(\cdot) \end{bmatrix}$$

where f is the activation function, as have already said a non-linear one. Precisely, the function used is the Logistic sigmoid function:

$$f(x) = \frac{1}{1+\exp(-x)}$$

References

[1] P. Arena, Luigi Fortuna, A. Gallo, S. Graziani and G. Muscato. In troduction motor modelling using multi-layer perceptrons. IEICE Trans fundamentals, E76-A (5) May 1993.

[2] R. P. Lippmann, An introduction to computing with neural nets. IEEE ASSP Magazine April 1987.

[3] G. Tzanateas, C.A.T. Salama and Y. P. Tsividis. A CMOS Bandgap voltage reference. IEEE journal of solid-state circuits, SC-14 (3), June 1979.

[4] Yannis P. Tsividis and Richard W. Ulmer. A CMOS voltage reference. IEEE journal of solid-state circuits, SC-13 (6), December 1978.

[5] Eric A. Vittoz and Olivier Neyroud. A low voltage CMOS Bandgap reference. IEEE journal of solid-state circuits, SC-14 (3), June 1979.

[6] Eric A. Vittoz and Jean Fellrath. CMOS analog integrated circuits based on weak inversion operation. IEEE journal of solid-state circuits, June 1977.

[7] DARPA. Neural Network Study. Lexington, MA: M.I.T. Lincoln Laboratory, 1988.

[8] Hagan, M. T., and M. Menhaj. Training feedforward networks with the Marquardt algorithm. IEEE Transactions on Neural Networks, 5 (6) 989–993, 1994.

[9] Hagan, M. T., H. B. Demuth, and M. H. Beale. Neural Network Design. Boston, MA: PWS Publishing, 1996.

[10] S. Graziani. Introduzione alle reti neurali. Dip. Elettrico, elettronico e sistemistico – Università degli studi di Catania.

Modelling The Bipolar Transistor Using Multibias S Parameters Sets

Tonio Biondi

DEES, Engineering Faculty, University of Catania
Viale A. Doria 6, 95125 Catania - Italy

Introduction

Modelling the bipolar transistor in the high frequency range is an involved problem that has been taken as subject of several works in literature, arousing the interest of many scientists. The reason is that several physical phenomena take place within the active region of the device and even if it's possible to accurately model the static or low frequency non-linear behaviour of the device the same can't be said when it's working in the high frequency range.

Moreover as bipolar technologies push towards the deep submicrometer features size it becomes more and more difficult to understand the influence of intrinsic and extrinsic parasitic capacitances and resistances on the high frequency performance of the device.

This is due to the need of taking into account distributed effects involved in the behaviour of the device. Another factor determining the complexity of the problem is the particular technological process involved in the fabrication of high frequency bipolar transistors. Indeed, in polysilicon emitter bipolar devices, a layer of thin oxide, a few nanometers thick, is incorporated, and can't be removed, at the polysilicon-silicon interface of the emitter.

This layer is responsible of increasing the base and emitter resistances, owing to reduced hole mobility in the polysilicon emitter; moreover, the current dependence of AC base and emitter resistances is substantially modified by the presence of this layer. In this paper two models will be presented, both of them are based on a particular parameter extraction technique that will be fully explained in the next. The reason of doing so is that this technique allows accurately prediction of the frequency and noise behaviour of the device within the forward active region.

A first purpose of this work is thus the creation of a model that can be successfully used for small signal simulations of devices operating in the forward active region at different bias points.

While the first of the two implemented models has the only purpose to provide a multibias small signal model suitable over the entire forward active region, the other is a first attempt of a large signal model that will be hopefully accomplished in the next future.

Extraction Technique of S. Rinaudo

It is well known from the literature that it's possible to describe the frequency behaviour of the bipolar transistor using a small signal model working at a fixed bias level.

The topology of the model can be a simple π or T one like in [1] and [2], but some authors ([3]-[12]) have included additional components to take into account parasitics effects when unde-embedded experimental data have to be used. Example of parasitic components are overlap capacitances in parallel with the external terminals, due to oxide layers used to passivize the device, inductances in series with the terminals, due to the metal strip used to interconnect the external pads of the probe station with the device.

Model parameters can be determined in various ways both calculated by solving a linear system, and optimized with a least square root algorithm if the number of equations is smaller respect to the unknowns of the system.

In literature this technique has been extensively used obtaining very good results in a frequency range exceeding the transition frequency of the device being Modelled. For the purposes of this paper a similar technique will be adopted. The model used is the well-known π-hybrid model of the bipolar transistor shown in Figure 1.

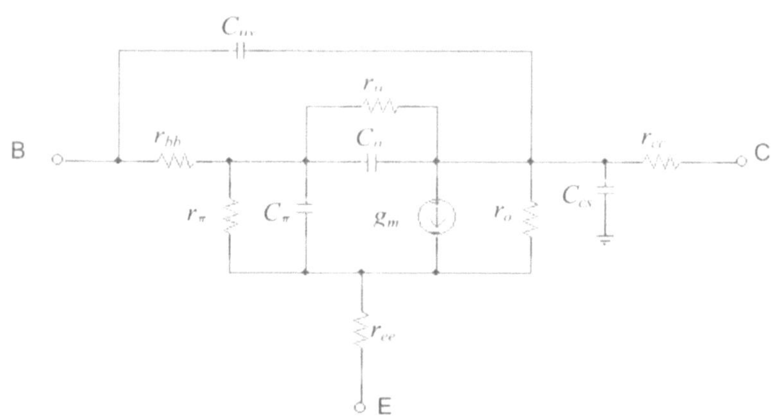

Figure 1. Equivalent π-hybrid model for the bipolar transistor, basic topology

Parameters are extracted by symbolical solution of the linear system obtained comparing the calculated and measured network parameters of the device at each measured frequency point.

This technique has been experimented by S. Rinaudo in STMicroelectronics, Catania site. Unfortunately, only eight of the eleven parameters of the topology shown before can be determined in a strictly analytical way because only four complex equations, and thus eight real ones, can be used, for each frequency point. In order to increase the number of equations, the basic relationships regulating the behaviour of the small signal model of the bipolar transistor have been added to the system.

These relationships, suitable for the input conductance and the transconductance of the bipolar transistor, coming from a first order analysis of the small signal model, are listed below:

$$g_{\pi} = \frac{I_B}{V_T} \tag{1}$$

$$g_m = \frac{I_C}{V_T} \tag{2}$$

where I_B and I_C are the measured DC base and collector currents. Parameters that can't still be analytically determined by direct solution of the system have to be fitted according to a least square root rule. The same is true for those components that can be added to the original topology to make the model more accurate in the high frequency range. This case goes for the non-zero collector-substrate conductance, the parallel capacitances and the series inductances. The finite collector-substrate conductance is an intrinsic effect of the device, indeed it models the substrate as a real conductor. Thus it has to be included in the model topology even if the experimental data have been accurately de-embedded. It was neglected in the basic topology presented before to simplify the model as usually is done, being the conductivity of the substrate very low. On the other hand the parallel capacitances and the series inductances come from the structure used to make possible the experimental measurements on the device, not from the device itself.

Their effects on the scattering parameters should have been quantified and subtracted from measured data in order to make these data representative of the device alone. So while the collector-substrate conductance has to be included in any case to make the model more accurate, the other components have to be added only if unde-embedded data are available. The complete topology containing all the components described above is shown in Figure 2.

At a fixed bias level, model parameters are calculated (or fitted as explained above) at each measured frequency point up to 10 GHz and stored in a database. The technique imposes an upper limit to the higher frequency at which model parameters can be extracted, in fact the model has to be approximately frequency independent and so the values extracted up to 10 GHz have to fit even at higher frequencies.

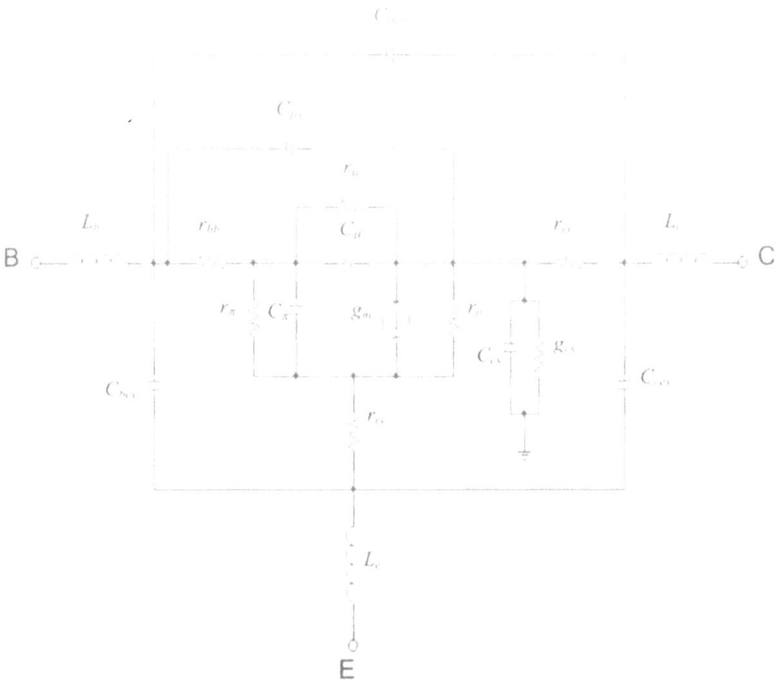

Figure 2. Equivalent π-hybrid model for the bipolar transistor, complete topology

If the measured data have been accurately de-embedded, this extraction technique provides nearly frequency independent parameters of the basic topology model shown in Figure 1, with the only exception of few measurement oscillations at very low frequencies.
So the frequency independent model can be determined averaging each model parameter over frequency or taking a random value in the flat part of the characteristics.
The frequency dependence of extracted model parameters for the basic topology is shown in Figures 3 to 13.

Figure 3. Frequency dependence of g_m

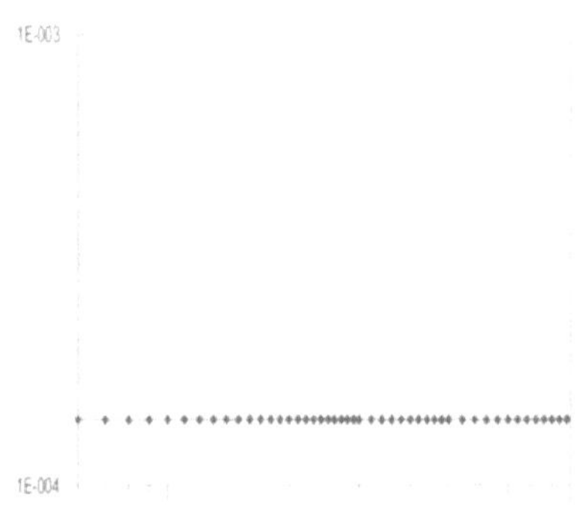

Figure 4. Frequency dependence of g_π

As shown in Figure 3 and 4, the transconductance g_m and the input conductance g_π are perfectly frequency independent, this is due to the fact that their values are constrained by equations (1) and (2), at each frequency point.

Actually these relationships give a starting point of the solution of the system not the solution itself, in fact they are equations of the system, like the other eight ones derived from the network parameter of the bipolar transistor, but in explicit form. The value of these parameters is thus strictly frequency independent if the symbolical solver converges, and this case goes for bias points within the forward active region of the device.

The same parameters show significant frequency dependence if the solution is to be found in a bias point outside the forward active region. In this case the symbolical solver doesn't converge and the value of these parameters can differ significantly from that predicted by equations (1) and (2) at certain frequency points.

While the frequency dependence of the feedback conductance g_μ can be neglected at frequencies below 5 GHz, as shown in Figure 5, it becomes stronger at higher frequencies when the capacitive impedance in parallel with it become too much small to allow a correct determination of that parameter. This fact is also demonstrated by the frequency dependence of the feedback capacitance C_μ shown in Figure 6. Its value is quite constant at frequencies higher than 3 GHz while become uncertain as frequency decreases, being dominant the contribute of the feedback conductance. The same is true for the output conductance g_o shown in Figure 7.

In this case the collector-substrate capacitive impedance dominates the high frequency behaviour as it can be seen in Figure 8 where the collector-substrate capacitance versus frequency is depicted. While being quite constant up to 3 GHz, the output conductance becomes frequency dependent at higher frequencies assuming negative values. In the same way the collector-substrate capacitance shows a hopping behaviour at lower frequencies becoming less sensitive as frequency goes over 4 GHz.

The general rule according to which capacitive elements are less frequency dependent in the high frequency range also applies to the base-emitter capacitance and to external base-collector capacitance, whose frequency dependence is shown in Figure 9 and 10.

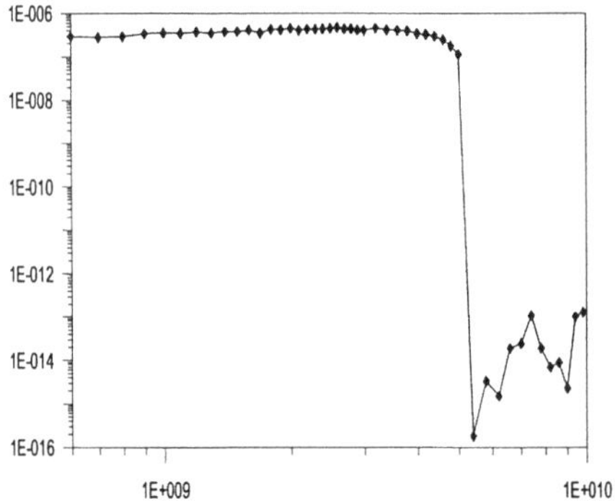

Figure 5. Frequency dependence of g_μ

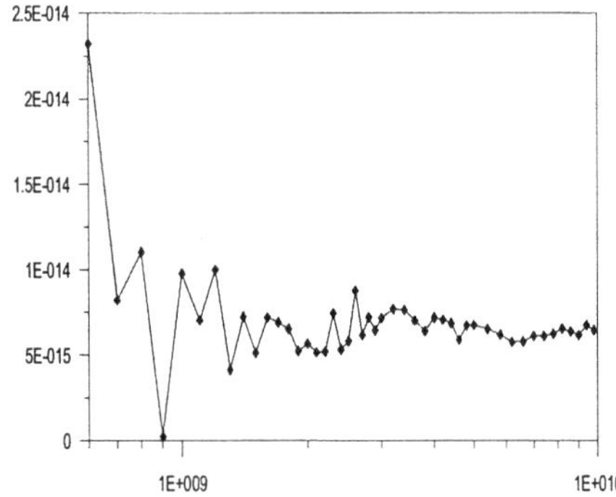

Figure 6. Frequency dependence of C_μ

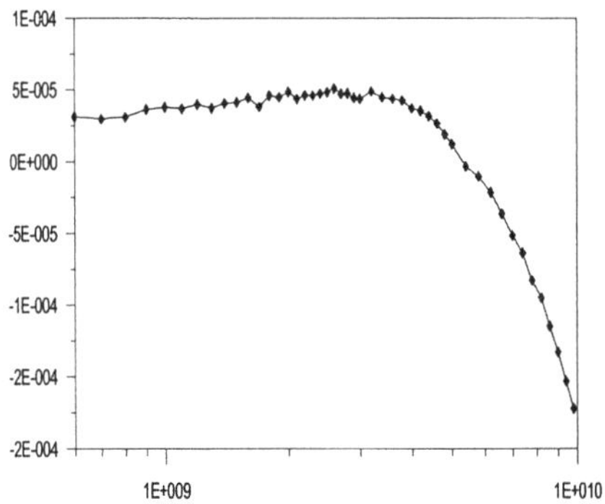

Figure 7. Frequency dependence of g_o

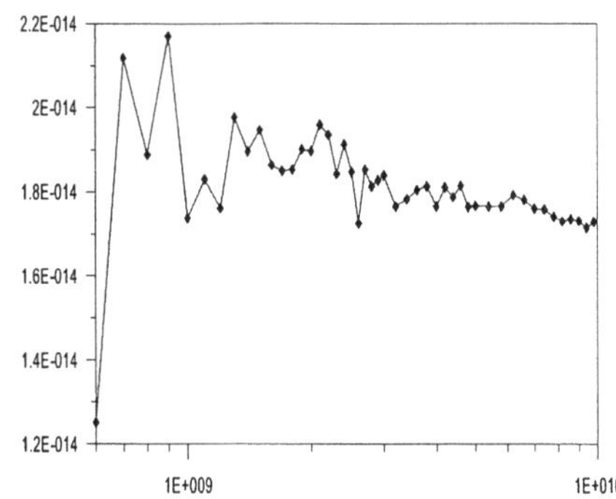

Figure 8. Frequency dependence of C_{cs}

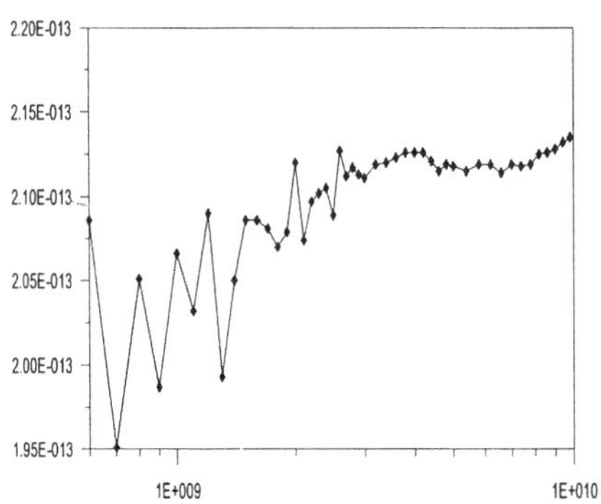

Figure 9. Frequency dependence of C_π

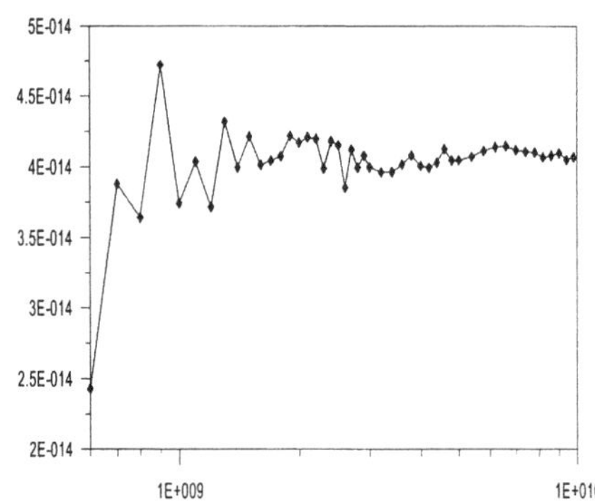

Figure 10. Frequency dependence of $C_{\mu x}$

Its value is quite frequency independent over 3 GHz when the impedance of this component become dominant respect to the one offered by the input conductance g_π in parallel with it. A quite different logic is to be done to justify the frequency behaviour of the three series resistances of the basic topology model.

They show weaker frequency dependence at higher frequencies being this dependence very low over the entire range of measured points, as it's demonstrated in Figure 11, 12 and 13.

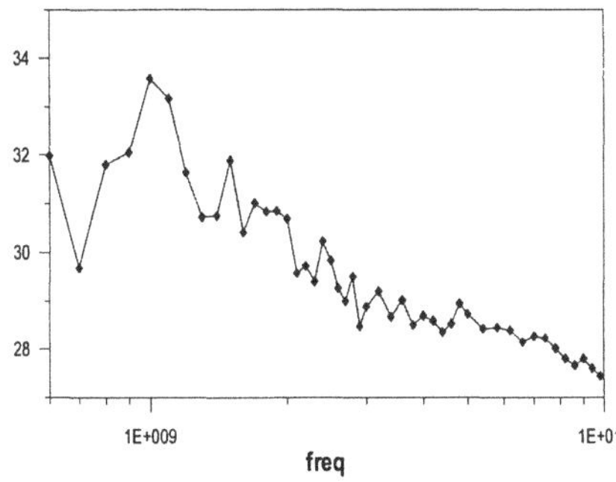

Figure 11. Frequency dependence of r_{cc}

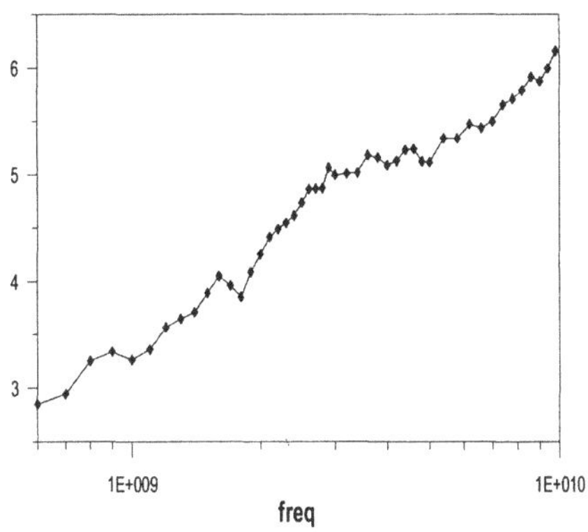

Figure 12. Frequency dependence of r_{ee}

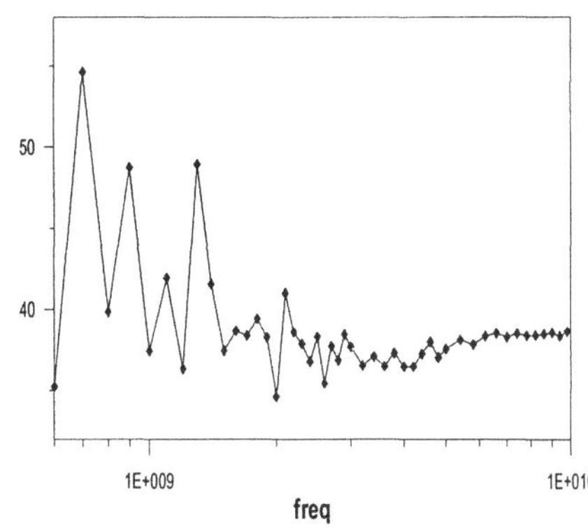

Figure 13. Frequency dependence of r_{bb}

For the base resistance this behaviour is due to the series impedance offered by the parallel connection of the base-emitter conductance and capacitance that influences more strongly the input impedance of the device at lower frequencies making the extraction of the base resistance more difficult. The same is true for the collector resistance whose value, at lower frequencies, is affected by series impedance offered by the collector-substrate capacitance and by the output conductance.

On the other hand the emitter resistance doesn't show significantly hopping values over the entire range of measured frequencies, but it doesn't reach a constant value becoming greater as frequency increases. It shows a clearly inductive behaviour, thus leading to think that a residual of parasitic inductance at the emitter pad is still affecting the experimental data even if a de-embedding procedure has been carried out. In this case the low frequency value has to be taken in order to minimize the effects of that parasitic.

This extraction technique has been fully validated both in frequency and noise performances ([1]).

The model of Figure 1, whose parameters have been randomly taken in the frequency range within they were extracted, shows very good agreement with measured data over the frequency range within it was extracted (0.1 to 10 GHz). The frequency behaviour of each extracted parameter explained before is thus entirely confirmed by experimental results. Even if some model parameter becomes frequency dependent at certain frequencies (conductances at higher frequencies, capacitances and series resistances at lower ones) it has the correct frequency independent value within the range in which it dominates the behaviour of the device. In this way simulated data reproduce quite perfectly measured ones both at lower and higher frequencies.

The technique is suitable in a range of bias levels covering the forward active region of the device. The topology has, in fact, to be modified if the model is working outside this region. In the cut-off region the device can be successfully Modelled with a π connection of capacitances representing the reverse-biased base-emitter, base-collector and collector-substrate junctions. The topology of Figure 1 proves unacceptable even in the high injection level region where the dynamic emitter crowding causes the base-emitter capacitance to be split toward the periphery of the device. An extra capacitance connected between the external base and the intrinsic emitter has to be added to the basic topology to take into account this effect; the even modified topology could be unsuitable owing to the intrinsic non-linear behaviour of the device working in this region. When high level injection takes place the performances of the model are in fact no longer acceptable since the frequency dependence of parameters becomes stronger. In this working condition there isn't a frequency range within which model parameters can be considered quite constant, this makes the extraction senseless because the entire model is frequency dependent and thus unacceptable. In such a non-linear condition a linear model proves inadequate even if a linear behaviour, such that described by scattering parameters, is to be performed.

Description of the Implemented Models

The extraction technique described in the previous section has been exploited to create two different models for high frequency bipolar transistors. The first, named tabular model, is a multibias small signal model that unifies the models extracted using the technique of S. Rinaudo at different bias levels within the forward active region according to an interpolating algorithm. The second is a macromodel obtained modifying the bias dependence of some components of the ST SPICE Gummel-Poon model in order to reproduce extracted data at different bias levels.

Tabular Model

In this section the most important features of a tabular model, based on the extraction technique explained before, will be treated. The aim is to give the designer an accurate high frequency model for small signal simulation at different bias levels within the forward active region of the device. This model can be successfully used for the design of circuits operating in that region such as low noise amplifiers. If all the devices contained in the circuit work in the forward active region, a small signal model could be extracted for each of them at the exact operating point at which it works. At each bias point a model reproducing the measured frequency and noise performances of the device in a very accurately way could be used. Each device contained in the circuit will be Modelled by a set of parameters expressly extracted, giving the best performances can be expected.

On the other hand the operating point of the devices contained in the circuit should be known before the extraction will be performed. Then for each particular bias point the scattering parameters of the device have to be measured and the model has to be extracted.

If an approximation of the extracted model can be accepted instead of the extracted model itself, it's possible to proceed to the extraction in some bias points within the linear region of the device interpolating the model between a measured bias point and the next. The performances of this model will be the same of the extracted one if the simulation is performed in a measured bias point, an error due to the interpolation will be made when the simulation is to be performed in bias points where the model hasn't been extracted. The performances of the interpolated model can be still acceptable if the measurement setup is suitably set and the interpolating algorithm is effective enough. The model thus obtained isn't an extension of the extracted small signal model, in the sense that it can be used only in a bias range within the extraction domain. This model can't give any further informations about the non-linear behaviour of the device since the validity of the extracted model it interpolates is limited to the linear active region.

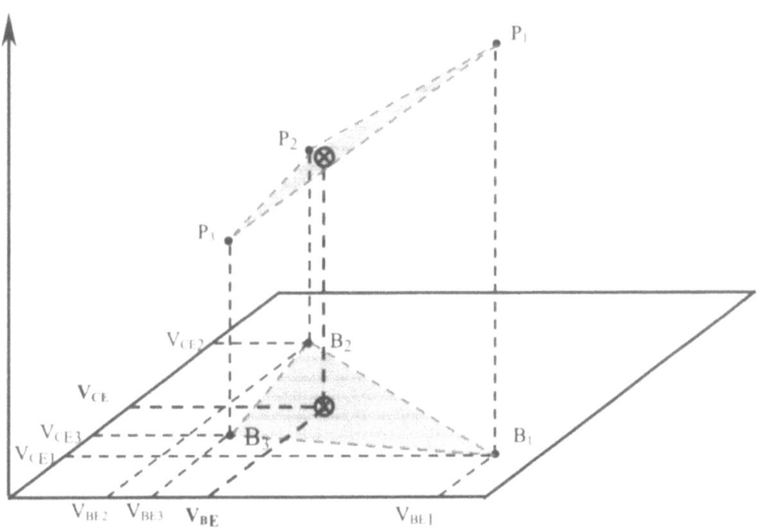

Figure 14. Interpolation technique for the tabular model

The usefulness of this kind of model is that it can be used for all the devices working in the circuit if it's designed to distinguish which set of model parameters is to be used according to the operating point being simulated. If the measurement setup is properly set, *i.e.*, with a large number of bias points within the operating regions of the devices, the performances of the tabular model over an interpolated point could be comparable to that of the extracted model.

The name tabular model comes from the consideration that all the useful informations the model needs to perform a simulation are stored in a tabular database. Each row of the database contains the description of the model at a fixed bias point in which it has been extracted, containing the DC voltages and currents and the value of the extracted small signal parameters.

In order to avoid problems connected with the frequency dependence of model parameters discussed in the previous section, parameters have to be extracted in the frequency range in which they show quite constant values for each bias point. So conductances have to be extracted at lower frequencies, and capacitances and resistances at higher frequencies.

As shown in Figure 14, a zero order interpolation has been used to calculate model parameters of the tabular model in a bias point in which the extraction hasn't been performed. First of all the three measured bias points (B1, B2 and B3) having smallest distance from the one which has to be simulated (B) are identified.

Two coordinates, the base-emitter voltage (VBE1, VBE2 and VBE3) and the collector-emitter voltage (VCE1, VCE2 and VCE3), identify each point, the bias domain being two-dimensional. The model parameter (P1, P2 and P3) corresponding to each one of these three points is found in the database.

Then the plane passing through these three parameters is calculated and the interpolated parameter (P) is obtained projecting the simulated bias point (B) on that plane.

The same is repeated for all the eleven parameters, obtaining the entire interpolated model. This algorithm is a simple extension of the linear interpolation technique suitable for one-dimensional domains to a two-dimensional one.

Once model parameters have been taken or interpolated from the database, depending on the bias point being simulated, the Y parameters of the topology showed in Figure 1 are calculated.

This can be done into three steps: calculating the Y parameters of the net obtained from Figure 1 neglecting the three series resistances; converting the Y parameters into Z parameters and adding the three series resistances; converting the Z parameters into Y parameters.

For the first of these three steps the following equations can be used:

$$Y_{11} = Y_{bc} + \frac{1}{r_b} \cdot \left(1 - \frac{1}{\Delta \cdot r_b}\right) \tag{3}$$

$$Y_{12} = -Y_{bc} - \frac{Y_{b'c}}{\Delta \cdot r_b} \tag{4}$$

$$Y_{21} = -Y_{bc} + \frac{g_m - Y_{b'c}}{\Delta \cdot r_b} \tag{5}$$

$$Y_{22} = g_{ce} + Y_{bc} + Y_{b'c} + \left(g_m - Y_{b'c}\right) \cdot \frac{Y_{b'c}}{\Delta} + Y_{CS} \tag{6}$$

In these equations $Y_{b'e} = j \cdot \omega \cdot C_\pi + \frac{1}{r_\pi}$ and $Y_{b'c} = j \cdot \omega \cdot C_\mu + \frac{1}{r_\mu}$ are the equivalent admittances of the base-emitter and base-collector branch, respectively; $Y_{bc} = j \cdot \omega \cdot C_{\mu x}$ and $Y_{cs} = j \cdot \omega \cdot C_{cs}$ are the external base-collector and the collector-substrate admittances respectively, $g_{ce} = \frac{1}{r_o}$ is the output conductance and $\Delta = \frac{1}{r_b} + Y_{b'c} + Y_{b'e}$.

Once the Y parameters of the entire net have been calculated the AC currents can be obtained using the well-known relationships listed below:

$$i_b = Y_{11} \cdot v_{be} + Y_{12} \cdot v_{ce} \tag{7}$$

$$i_c = Y_{21} \cdot v_{be} + Y_{22} \cdot v_{ce} \tag{8}$$

where i_b, i_c v_{be} and v_{ce} are the AC values of the terminal currents and voltages of the net.

Macromodel

An idea similar to the one used for the tabular model can still be used to extend the bias range covered by the multibias small signal model extracted using the technique of S. Rinaudo discussed above. In this case equations describing the bias dependence of the components of the model have to be implemented, instead of an interpolating algorithm, in order to work with continuos function that can be integrated to obtain the dynamic large signal behaviour of the model. The purpose is a model behaving like the small signal one discussed above, in the forward active region, and still having good performances, in terms of scattering parameters, even in bias points outside this region. So it can be used like a multibias small signal model whose bias domain is enlarged respect to the one of the tabular model discussed in the previous section. Circuits working in linear region such as low noise amplifiers, in which the bias point is chosen to ensure the minimum noise figure, can successfully work in this range of bias levels.

The performances of this kind of model in the dynamic large signal behaviour, including the high injection, saturation, quasi-saturation and cut-off region, could be then optimized using a multibias small signal model fitting the measured scattering parameters in these regions.

The topology of such a model in the cut-off region is already known in fact a π connection of capacitances Modelling the reverse-biased junction of the device has been successfully used in literature with this aim.

On the opposite the topology of a small signal model that fits the scattering parameters in the high injection, saturation and quasi-saturation regions is still unknown.

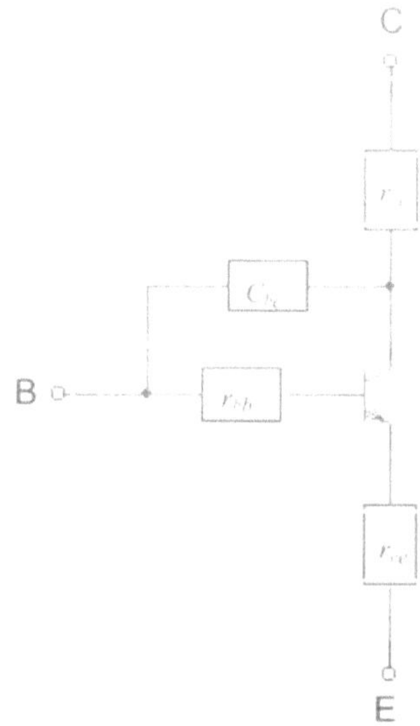

Figure 15. Schematic view of the macromodel

Actually this is a limit for the range of bias levels that can be covered by a multibias small signal model and also for the working regions that can be performed by a large signal model. Moreover these regions are the most important to be Modelled if a strongly non-linear circuit like a power amplifier is to be designed.

In the next of this section it will be explained what it can be done to pursue the first of the aims explained above, trying to extend the bias domain of the tabular model using large signal equations for the bias dependence of the components of the model. The implemented macromodel including the intrinsic bipolar transistor, the three series resistances and the external base-collector capacitance is shown in Figure 15.

Initially the bias dependence of each model parameter has been Modelled using equations similar to the ones describing the behaviour of the Gummel-Poon model, using the ST SPICE version of the Gummel-Poon ([13], [14], [15]) which adds to the basic equations of that model some parameters Modelling the base pushout effect. These equations will be described in the next.

While most of the bias dependent model parameters can be successfully fitted with the ST SPICE Gummel-Poon equations even when high injection effect occurs, other parameters show a strange behaviour that can't be predicted using that model. The latter case goes for the base-collector conductance, the collector-substrate capacitance, and the base and emitter series resistances. Owing to this fact some equations have to be modified in order to provide a good fitting with experimental data, while other equations still showing a significant mismatch haven't been changed for the reasons that will be explained in the next.

The case of the base-collector conductance is treated first of all because it can be solved immediately. Indeed at the working frequencies the impedance of the base-collector capacitance is much smaller than the base-collector resistance thus it can be very probable that the accuracy of extracted data isn't satisfactory. The symbolic solver, in fact, stops the iterations and find out the solution of the system when the accuracy of extracted data is higher than a fixed tolerance. Being the base-collector conductance smaller than the base-collector capacitive admittance at every bias point, a significant error could have been done extracting that parameter, even if the accuracy of the global solution is still below the limit imposed by the tolerance.

So even if the bias dependent base-collector conductance Modelled with ST SPICE Gummel-Poon equation doesn't fit extracted data, and there is still a significant mismatch between simulated and extracted values of that component, it doesn't cause any real problem being very low the impact of the base-collector conductance on S parameters, as it will be confirmed in the next.

Also the extracted base-collector conductance provides a considerable mismatch even with DC current measurements in reverse region, for these reasons neither the Gummel-Poon parameters nor the implemented equation haven been changed in order to fit the extracted data. It has been used the same equations of ST SPICE Gummel-Poon with parameters extracted by reverse Gummel plot fitting.

The case of the collector-substrate capacitance is a bit more difficult and actually it hasn't been solved, being connected to the parasitic pnp bipolar transistor that hasn't been considered in this analysis. The effects of this component on S parameters can't be neglected, this capacitance being important in the output matching and affecting the magnitude and phase of S_{22} at higher frequencies. So a mismatch between simulated and measured S_{22} has to be expected at higher frequencies.

Different consideration can be made for the bias dependence of AC base and emitter series resistances.

Their behaviour is strongly influenced by the particular technology used to fabricate self-aligned high frequency bipolar transistors. Indeed a thin oxide layer grows at the polysilicon-silicon interface of the emitter in such devices, which strongly influences both the frequency and bias behaviour of the base-emitter junction.

An accurate analysis of the physical phenomenon is thus needed in order to model the behaviour of that part of the device in a suitable way.

By the light of these considerations, the different parts of the model have been Modelled with the equations that fitted experimental data in the best way.

Some of these equations, those Modelling the intrinsic device, are the same as those implemented in the ST SPICE Gummel-Poon model, thus including the base pushout parameters.

Equations for the base, emitter and collector series resistances have been instead modified, the ones implemented in the ST SPICE Gummel-Poon model proving inadequate to describe the AC behaviour of these components at different bias levels. The static part of the intrinsic model includes the base-emitter conductance (g_π), the base-collector conductance (g_μ), the transconductance (g_m) and the output conductance (g_o).

Each of these elements is bias dependent according to the same equations implemented in the ST SPICE Gummel-Poon model listed below:

$$g_\pi = \frac{g_{mF}}{BF} + g_{lF} \tag{9}$$

$$g_{\mu} = \frac{g_{mR}}{BR} + g_{lR} \tag{10}$$

$$g_{o} = \frac{g_{mR}}{q_b} \cdot \left[1 + \frac{q_1 \cdot (I_F - I_R)}{q_b \cdot q_3 \cdot IKR}\right] + \frac{q_1 \cdot (I_F - I_R)}{q_b \cdot VAF} \tag{11}$$

$$g_{m} = \frac{g_{mF}}{q_b} \cdot \left[1 - \frac{q_1 \cdot (I_F - I_R)}{q_b \cdot q_3 \cdot IKF}\right] - \frac{q_1 \cdot (I_F - I_R)}{q_b \cdot VAR} - g_o \tag{12}$$

In these equations g_{mF}, g_{mR}, g_{lF}, g_{lR}, I_F, I_R, q_1, q_2, q_3 and q_b are defined as follow:

$$g_{mF} = \frac{\partial I_F}{\partial V_{be}} = \frac{IS}{NF \cdot V_T} \cdot e^{\frac{V_{be}}{NF \cdot V_T}} + g_{\min} \tag{13}$$

$$g_{mR} = \frac{\partial I_R}{\partial V_{bc}} = \frac{IS}{NR \cdot V_T} \cdot e^{\frac{V_{bc}}{NR \cdot V_T}} + g_{\min} \tag{14}$$

$$g_{lF} = \frac{\partial I_{LF}}{\partial V_{be}} = \frac{ISE}{NE \cdot V_T} \cdot e^{\frac{V_{be}}{NE \cdot V_T}} \tag{15}$$

$$g_{lR} = \frac{\partial I_{LR}}{\partial V_{bc}} = \frac{ISC}{NC \cdot V_T} \cdot e^{\frac{V_{bc}}{NC \cdot V_T}} \tag{16}$$

$$I_F = IS \cdot \left(e^{\frac{V_{be}}{NF \cdot V_T}} - 1 \right) + g_{\min} \cdot V_{be} \tag{17}$$

$$I_R = ISN \cdot \left(e^{\frac{V_{bc}}{NR \cdot V_T}} - 1 \right) + g_{\min} \cdot V_{bc} \tag{18}$$

$$q_1 = \left(1 - \frac{V_{be}}{VAR} - \frac{V_{bc}}{VAF} \right)^{-1} \tag{19}$$

$$q_2 = \left(B \cdot \frac{I_F}{IKF} + \frac{I_R}{IKR} \right) \tag{20}$$

$$q_3 = \sqrt{1 + 4 \cdot q_2} \tag{21}$$

$$q_b = \frac{1}{2} \cdot q_1 \cdot [1 + q_3] \tag{22}$$

(equation parameters are written in full uppercase letters to distinguish them from symbolically defined current such as I_F).

These equations, with the exception of the one for g_{μ} shown in Figure 18, fit extracted data up to bias levels at which high injection occurs as it's shown in Figure 16, 17 and 19.

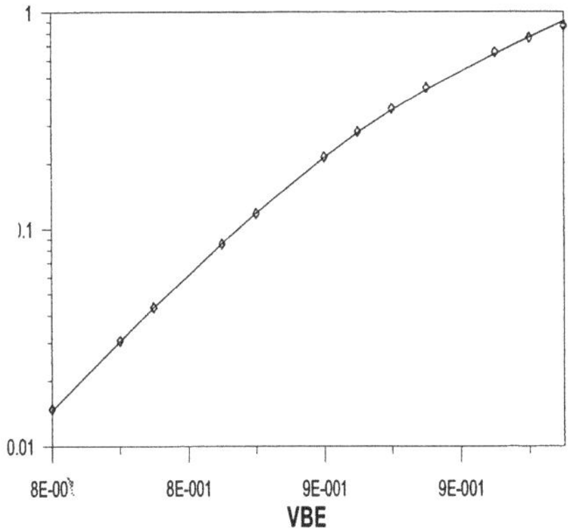

Figure 16. Comparison between simulated and measured g_m

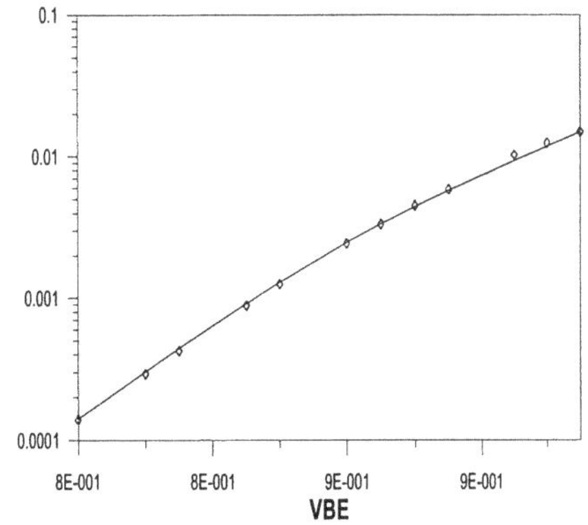

Figure 17. Comparison between simulated and measured g_π

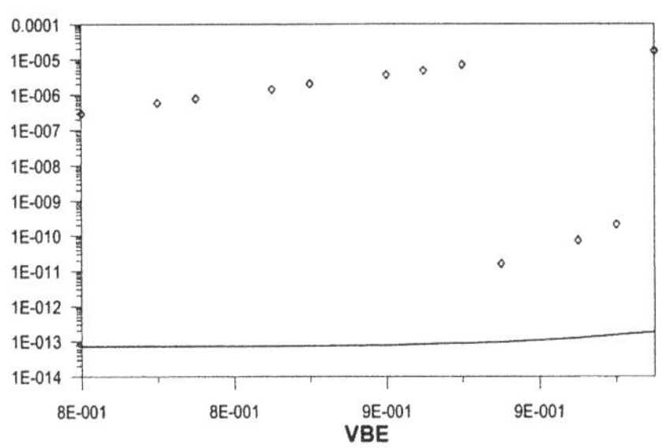

Figure 18. Comparison between simulated and measured g_μ

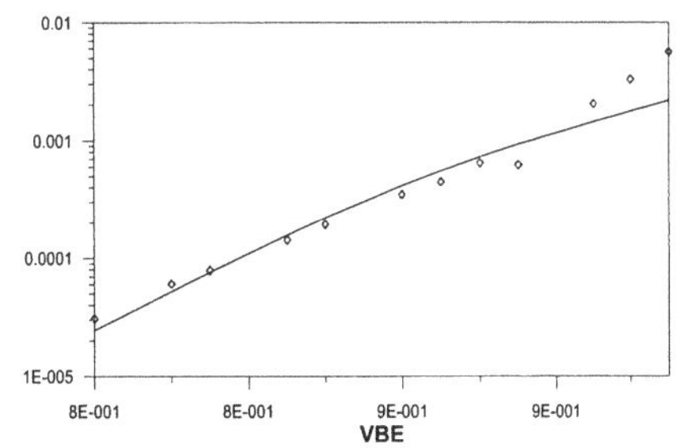

Figure 19. Comparison between simulated and measured g_o

This is possible thanks to the base pushout factor B (only present in ST SPICE Gummel-Poon model) that increases the charge storage in the base thus reducing the effective collector current when the base pushout (Kirk effect) occurs. The base pushout factor is given by

$$B = \left[1 + RW \cdot \frac{\left(\sqrt{(I_{CN} - I_O)^2 + RP} + (I_{CN} - I_O) \right)^{ENP}}{4 \cdot \left(I_{CN}^{\ 2} + RP \right)} \right]^2 \tag{23}$$

where $I_{CN} = \dfrac{I_C}{IKF}$ and $I_O = \dfrac{VJJ - V_{bc}}{VRP}$, while RW, RP, VJJ, VRP and ENP are the base pushout parameters.

Equations for the DC base, emitter and collector currents are listed below:

$$I_C = \frac{I_F - I_R}{q_b} - I_{BR} - I_{LR} \tag{24}$$

$$I_B = I_{BF} + I_{LF} + I_{BR} + I_{LR} \tag{25}$$

$$I_E = -(I_B + I_C) = -\frac{I_F - I_R}{q_b} - I_{BF} - I_{LF} \tag{26}$$

where I_{BF}, I_{BR}, I_{LF} and I_{LR} are defined as follows:

$$I_{BF} = \frac{I_F}{BF} = \frac{IS}{BF} \cdot \left(e^{\frac{V_{be}}{NF \cdot V_T}} - 1 \right) + \frac{g_{\min} \cdot V_{be}}{BF} \tag{27}$$

$$I_{BR} = \frac{I_R}{BR} = \frac{ISN}{BR} \cdot \left(e^{\frac{V_{bc}}{NR \cdot V_T}} - 1 \right) + \frac{g_{\min} \cdot V_{bc}}{BR} \tag{28}$$

$$I_{LF} = ISE \cdot \left(e^{\frac{V_{be}}{NE \cdot V_T}} - 1 \right) \tag{29}$$

$$I_{LR} = ISC \cdot \left(e^{\frac{V_{bc}}{NR \cdot V_T}} - 1 \right) \tag{30}$$

The dynamic part of the intrinsic model includes: the base-emitter capacitance (C_π), whose equations for the reverse (31) and forward (32) bias are listed below:

$$C_\pi = CJE \cdot \left(1 - \frac{V_{be}}{VJE} \right)^{-MJE} \tag{31}$$

$$\begin{aligned} C_\pi = {} & \frac{CJE}{(1 - FC)^{(1+MJE)}} \cdot \left(1 - FC \cdot (1 + MJE) + \frac{MJE \cdot V_{be}}{VJE} \right) \\ & + \frac{g_{mF} \cdot TF}{q_b} \cdot \left[1 + XTF \cdot \left(\frac{I_F}{I_F + ITF} \right)^2 \cdot \left(\frac{I_F + 3 \cdot ITF}{I_F + ITF} \right) \cdot e^{\left(\frac{V_{bc}}{1.44 \cdot VTF} \right)} \right] \\ & - \frac{T_{FF} \cdot I_F}{q_b^{\,2}} \cdot q_1 \cdot \left(\frac{q_b}{VAR} + \frac{g_{mF}}{q_3 \cdot IKF} \right); \end{aligned} \tag{32}$$

the base-collector capacitance (C_μ), for reverse (33) and forward (34) bias:

$$C_\mu = XCJC \cdot CJC \cdot \left(1 - \frac{V_{bc}}{VJC} \right)^{-MJC} \tag{33}$$

$$C_\mu = XCJC \cdot \frac{CJC}{(1 - FC)^{(1+MJC)}} \cdot \left(1 - FC \cdot (1 + MJC) + \frac{MJC \cdot Vbc}{VJC} \right) + g_{mR} \cdot TR \tag{34}$$

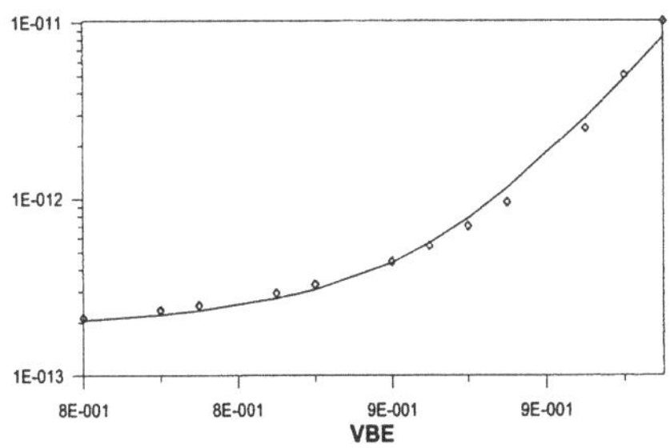

Figure 20. Comparison between simulated and measured C_π

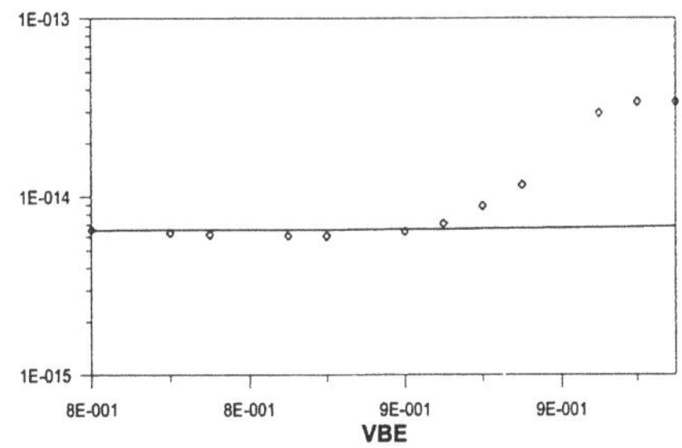

Figure 21. Comparison between simulated and measured C_μ

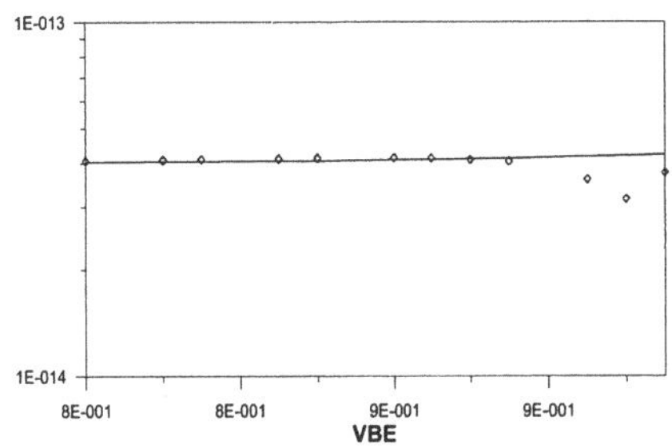

Figure 22. Comparison between simulated and measured $C_{\mu x}$

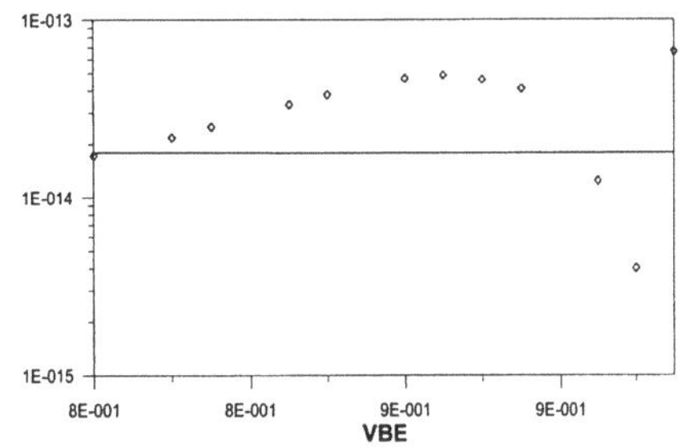

Figure 23. Comparison between simulated and measured C_{cs}

the extrinsic base-collector capacitance ($C_{\mu x}$), for reverse (35) and forward (36) bias:

$$C_{\mu x} = (1 - XCJC) \cdot CJC \cdot \left(1 - \frac{V_{Bc}}{VJC}\right)^{-MJC} \tag{35}$$

$$C_{\mu x} = (1 - XCJC) \cdot \frac{CJC}{(1 - FC)^{(1+MJC)}} \cdot \left(1 - FC \cdot (1 + MJC) + \frac{MJC \cdot VBc}{VJC}\right) \tag{36}$$

the collector-substrate capacitance (C_{cs}), for reverse (37) and forward (38) bias:

$$C_{cs} = CJS \cdot \left(1 - \frac{V_{cS}}{VJS}\right)^{-MJS} \tag{37}$$

$$C_{cs} = CJS \cdot \left(1 + \frac{MJS \cdot VcS}{VJS}\right) \tag{38}$$

The remaining components of the model are the three series resistances. To obtain the best fitting on multibias high-frequency S parameters it's of most importance to accurately predict the bias dependence of base, emitter and collector series resistances, these components (base and emitter resistances in particular) having a dominant role in the high-frequency behaviour of the device.

In order to fit the AC bias dependent collector resistance, ST SPICE Gummel-Poon equations can still be used. In fact it decreases monotonically with collector current as the base pushout modulated collector resistance of the ST SPICE Gummel-Poon model predicts. Nevertheless this equation can bring to DC convergence problems owing to the fact that it's strictly related to base pushout parameters, responsible of DC collector current in high injection region. For this reason it has been preferred to split the collector resistance into two parts: the DC resistance being bias independent and the AC resistance being bias dependent according to the extracted data. The AC part of the collector resistance was Modelled with a second order polynomial equation introducing the three coefficient of the polynomial as fitting parameters.

This equation is reported below:

$$rcc = RC0 - RC1 \cdot \left(\frac{I_C}{IKF} \right) - RC2 \cdot \left(\frac{I_C}{IKF} \right)^2 \tag{39}$$

where RC0 is the low current level collector resistance, RC1 and RC2 are fitting parameters of the polynomial. The collector current is normalised with the same parameter as in the ST SPICE Gummel-Poon model (IKF, the forward beta roll-off knee current) to reduce the number of fitting parameters. However a new parameter could have been introduced to improve the performances as it was done in the base and emitter resistances equations described in the next. In Figure 24 a comparison between simulated and extracted data is depicted.

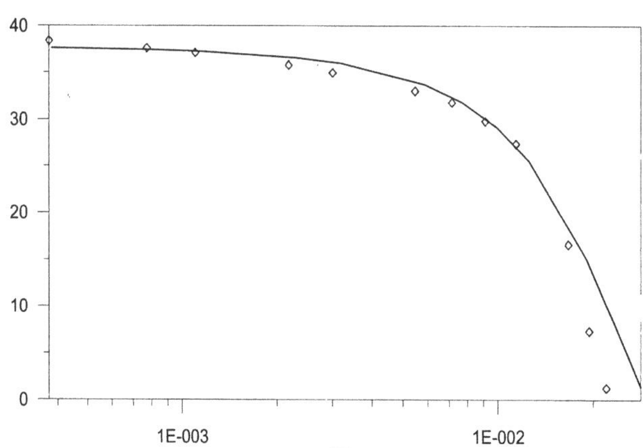

Figure 24. Comparison between simulated and measured collector resistance

The modified model also includes the effects of bias dependent AC base and emitter resistances. Like the collector resistance the DC value of these resistances is current independent in order to prevent convergence problems in DC and transient analysis, while assuring to fit the DC current measurements. The AC value of these resistances is current dependent according to the prediction of the extracted model. Both base and emitter resistances are Modelled with a pseudo-polynomial equation, the one for the base resistance contain a logarithmic term to ensure good results at low current levels while the one for the emitter resistance contains an hyperbolic term for the same reason. Simulated and extracted base and emitter AC resistances are compared in Figure 25 and 26.

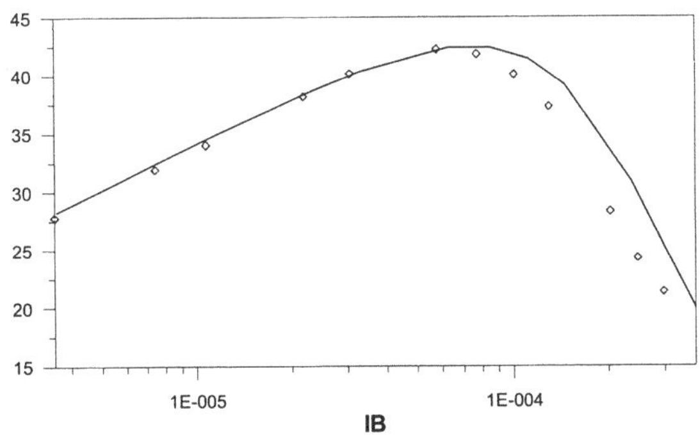

Figure 25. Comparison between simulated and measured base resistance

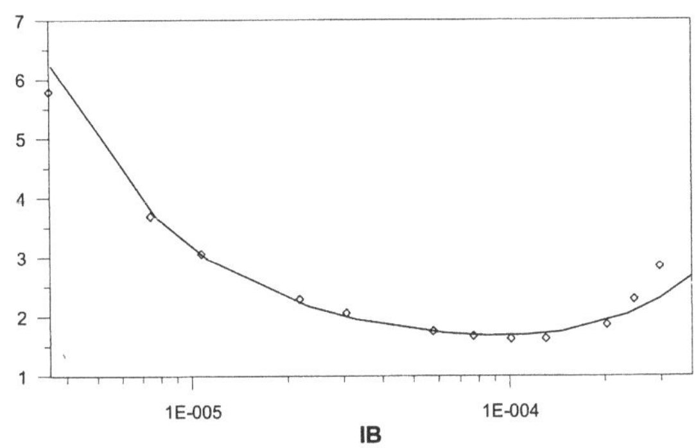

Figure 26. Comparison between simulated and measured emitter resistance

The equation for the base resistances is listed below:

$$rbb = RB0 + RBL \cdot \log\left(\frac{I_B}{IRBMAX}\right) + RB1 \cdot \left(\frac{I_B}{IRBMAX}\right) + RB2 \cdot \left(\frac{I_B}{IRBMAX}\right)^2 + RB3 \cdot \left(\frac{I_B}{IRBMAX}\right)^3 \quad (40)$$

where RB0 is the low current level base resistance, RBL is the slope of the logarithmic term at low base current level, IRBMAX is the base current value at which the resistance reaches its maximum value, RB1, RB2, RB3, are fitting parameters of the polynomial.

The equation for the emitter resistance is reported below:

$$ree = RE0 + REIP \cdot \left(\frac{I_B}{IREMIN}\right)^{-1} + RE1 \cdot \left(\frac{I_B}{IREMIN}\right) + RE2 \cdot \left(\frac{I_B}{IREMIN}\right)^2 \quad (41)$$

where RE0 is the low base current emitter resistance, REIP is the slope of the hyperbolic term, IREMIN is the base current at which the emitter resistance reaches its minimum value, RE1 and RE2 are fitting parameters of the polynomial. It should be noted that IRBMAX and IREMIN, normalizing the base current in the base and emitter resistance equations respectively, could be the same for both equations. In fact the current at which the base resistance reaches its maximum value is the same of the one at which the emitter resistance reaches its minimum value. This could be expected, in fact the physical phenomenon causing the base and emitter resistance modulation is the same: the presence of an oxide layer at the polysilicon-silicon interface in the emitter. Even if the consequences that this oxide layer have on the current dependence of base and emitter resistances are not so clear, and discovering them is beyond the scope of this paper, they influence base and emitter in the same manner. The optimized values of these parameters are in fact very similar to each other confirming the theory explained before.

Numerical Tools

The tabular model and the macromodel described in the previous section have been characterized and implemented thanks to two numerical tools whose features will be explained in this section.

S. Rinaudo, who also has experimented the extraction technique discussed before, created the first of these tools, named Toolsmg, in STMicroelectronics, Catania site. The other, named HDL-A, is a commercial tool developed by Mentor Graphics; it goes with ELDO, a simulation tool extensively used in STMicroelectronics for analog and mixed signal simulations.

Toolsmg by S. Rinaudo

Toolsmg is a user-friendly tool for full characterization of small signal equivalent circuit of integrated devices. It's suitable not only for bipolar transistors but also for MOS transistors and inductors, providing different modules allowing parameters extraction, network and noise parameters simulation, optimization and pole-zero analysis. The topologies used to model the bipolar transistor are the ones shown in Figures 1 and 2 and described in the section dedicated to the tabular model.

The most important feature of this tool is probably the one allowing the creation of a database in which are stored all the useful informations that can be extracted from network parameters of the device. The informations stored in the database come from the symbolic solution of the system obtained equating the measured network parameters of the device with the calculated ones for each measured frequency point up to 10 GHz. The solution involves usually a large number of iterations owing to the fact that the unknowns of the system are usually more than the equations, so an optimization has to be performed while the symbolical solver runs. Moreover symbolical solving and optimization have to be repeated at each measured frequency point leading to further delay. Nevertheless at a fixed bias point the database can be extracted in few minutes, the extraction time depending on the goodness of measured network parameters and on the de-embedding procedure that has been applied. It's also possible to use unde-embedded data, in this case parasitics are extracted internally during the solution of the system. After the database has been filled, the informations stored in it can be used to perform every kind of small signal analysis; alternatively it's possible to refine the values of extracted model parameters optimizing them in order to fit measured network parameters. An output ASCII file will be produced containing the more suitable device representation according to the choice of the user. Various types of representations of the device can be stored in this file including the parameters of a π-hybrid model like the one shown in Figure 1 or 2, the extracted network parameters (in the S, Y, Z, H, G, T or ABCD form), the noise parameters and a lot of other which are not mentioned for briefness' sake. Another important feature of this tool is that it can be easily interfaced with low level languages like C++ in order to automate the extraction procedure when multibias network parameters are available. This feature has been exploited for the purposes of this paper in order to provide full characterization of the bipolar transistor in term of multibias small signal parameters in the forward active region.

Mentor Graphics HDL-A

HDL-A is a powerful tool for behavioural Modelling of both analog and digital hardware. It is based on VHDL semantic and syntax but is optimized for analog description, in fact the syntax for analog hardware has been extended, while that of digital hardware has been restricted and simplified. It's designed to work both with analog (SPICE) and digital (VHDL) simulators. The complete tool including VHDeLDO simulator and HDL-A allows the Modelling and simulation of a general mixed-signal circuit with SPICE like syntax. The most important features of this tool will be briefly explained in the next. Each HDL-A model is connected to the external world by pins (ports for digital models), that make possible the communication between the model and all other elements composing the circuit, such as other HDL-A models, subcircuits or predefined components. Voltage and current at each pin are referred as across and through quantities respectively, they are the only quantities the model can share with the outer electrical world, *i.e.*, the rest of the circuit. The large signal behaviour of an analog model can be described in terms of ordinary differential equations, being an analog component a continuos-time and continuos-value system; VHDeLDO simulator, linearizing the large signal behaviour around the operating point, automatically calculate the small signal behaviour of the model. Alternatively, a small signal model can be explicitly described even if no large signal behaviour has been previously Modelled.

HDL-A allows the description of different behaviours for each kind of analysis performed by the simulator, thus the AC behaviour of the model could be different from the DC one or the TRANSIENT one. The behaviour can also be expressed in terms of equations in implicit form with an arbitrary number of unknowns, in this case an internal solver provides the solution of the system, that may be linear or not. An important feature of HDL-A is the C interface that gives the opportunity to describe the behaviour of the model in C/C++ language. The C interface makes also possible to include complex algorithms, formulas and every kind of user defined functions, implemented in C/C++, in the description of the model, thus expanding the descriptive possibility of the HDL-A language. The HDL-A model may be compiled after the C/C++ source code has been successfully written and compiled using the analog model compiler (amc) provided by the tool. Every C/C++ function has to be correctly declared inside each entity using that function in the HDL-A source code. Finally the complete path of all the C/C++ functions used in the HDL-A code has to be included in the netlist file to allow the simulator to find the executables files during the simulation. When VHDeLDO performs a simulation it doesn't know anything about the behaviour of the model, that appears like a black box sharing with the external world only the voltages and currents at the pins. So it has to find a solution of the net imposing an input to the model, may be a voltage or a current, and waiting for the output fed back by the model. The solution is found after some iteration whose number depends on the complexity of the net to be solved. It's worth noting that before the steady state solution is reached, during the transient, voltages in the nets and currents in the branches of the circuit could have unexpected values both positive and negative, thus the design of the model has to be robust in order to prevent convergence problems. This case goes in particular for the bipolar transistor during DC analysis, and thus during all the analyses that need a DC to be previously performed. The voltages at the external pins (*i.e.*, the voltages at the external terminals of the junctions) are provided and the currents (generated by forward or reverse biased junctions) flowing through these pins have to be found. Owing to the presence of the base, emitter and collector series resistances, the internal voltages biasing each junction of the transistor are different from the external ones, and this difference is the voltage drop across each of the three series resistances mentioned above. This voltage drop depends on the current flowing through the pins, so this situation creates a loop whose solution can be found by iteration. At low current level the voltage drop across the resistances is little and so the solution is found with few iterations, being quite equal the internal and external voltages. On the other hand, at high current level, when this voltage drop became a significant part of the external voltage, this loop can bring to a large number of iterations and in some cases the loop can't be solved owing to convergence problems.

The solution of the loop can be more and more difficult to find if the value of each of these resistances is high and depending on the current flowing through it, which is the case of the bipolar transistor. This is probably the most difficult problem to solve when a compact model for the bipolar transistor is to be implemented, making a macromodel a more suitable choice.

Implementation and Experimental Results

Both the tabular model and the macromodel have been implemented in HDL-A using the C/C++ interface. Only the strictly behavioural description part of the model has been written in HDL-A while the powerful C++ syntax has been used to define the complex formulas and algorithms setting up the structure of the model. As it was said in the previous section, the model has to be robust in order to provide correct values during the transient in which the simulator try to solve the net. During this transient voltages and currents are forced to have both positive and negative values thus the formulas and algorithms implemented in C++ have to be designed to prevent invalid floating point operations, such as division by zero or negative values of the arguments of logarithmic functions. The extraction of parameters of the small signal model at different bias points preceded the implementation of the model itself. The small signal model has thus been extracted: thirty-nine bias points were available with thirteen values of base-emitter voltage from 0.8 V to 0.97 V and three values of collector-base voltage from 0 V to 1.5 V.

For each bias point the eleven parameters of the small signal model shown in Figure 1 together with the DC voltages and currents have been stored in a database file. This file will be used to perform the interpolation for the tabular model and the curve fitting for the macromodel.

Tabular Model

For the tabular model the C++ source code contains both the interpolating algorithm and the equations (3), (4), (5) and (6) discussed in the section dedicated to the description of that model.

The HDL-A source code is divided into two parts: one describing the DC behaviour of the model and another describing the AC behaviour; this distinction is necessary to reduce the simulation time as will be explained. The model hasn't, in fact, to perform a DC analysis, calculating the bias point and linearizing around it to simulate the network parameters, the extracted small signal parameters being yet stored in the database file. It has only to sense the bias point to be simulated to allow a correct interpolation of the small signal parameters if the bias point doesn't correspond to a measured one. Separating the DC part of the behavioural description is important to reduce significantly the simulation time, in fact the interpolating algorithm runs whenever the analysis in which it's called has to be performed and it can't be otherwise. So if the DC and AC behaviour would have been described together the database file will be read and the algorithm will be performed at each DC and AC iteration, leading to a too much long simulation time. In order to reduce the simulation time small signal parameters are calculated only during the DC analysis, then they are stored in static variables, thus avoiding the loss of information between a call of the C++ function and the next, when the AC analysis will be performed. Parameters are thus calculated only once during the DC analysis and then used at each frequency point during the AC analysis.

The same has been done for the macromodel implementation, but in this case the small signal parameters are calculated after the bias point of the device has been found during the DC analysis being this model independent from the database file. Both the tabular and the macromodel, implemented in HDL-A according to this rule, showed simulation time for the AC analysis comparable to that of a standard model such as the ST SPICE Gummel-Poon model. As it has been said before, the bias range that can be successfully Modelled with a small signal model extracted according to the technique of S. Rinaudo is bounded to the forward active region of the device. For the measured bias points within this region comparisons between simulated and measured S parameters show impressive performances of the model even beyond the frequency range within which parameters have been extracted.

Such a comparison for the tabular model is shown in figures from 27 to 34, corresponding to eight different values of the collector current at zero-bias for the base-collector junction.

The scattering parameters have been simulated from 0.3 GHz up to 50 GHz, model parameters being extracted up to 10 GHz.

Over the entire frequency range the error between simulation and measurement is very small confirming the effectiveness both of the extraction technique and the choice of taking the value of conductances al lower frequencies and that of capacitances and resistances at higher frequencies.

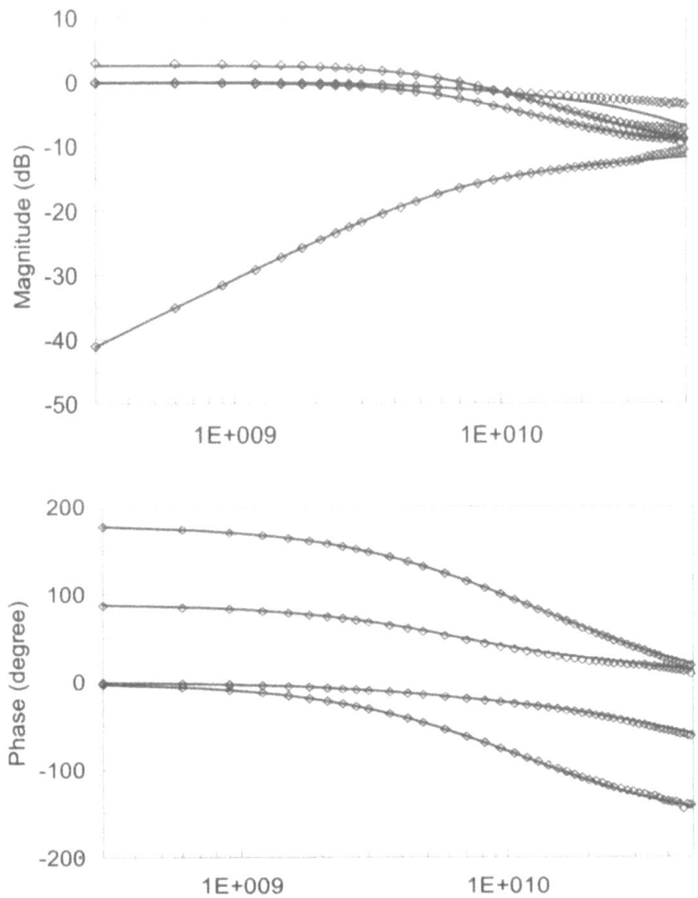

Figure 27. Comparison between simulated and measured S parameters at I_c=0.4 mA for the tabular model

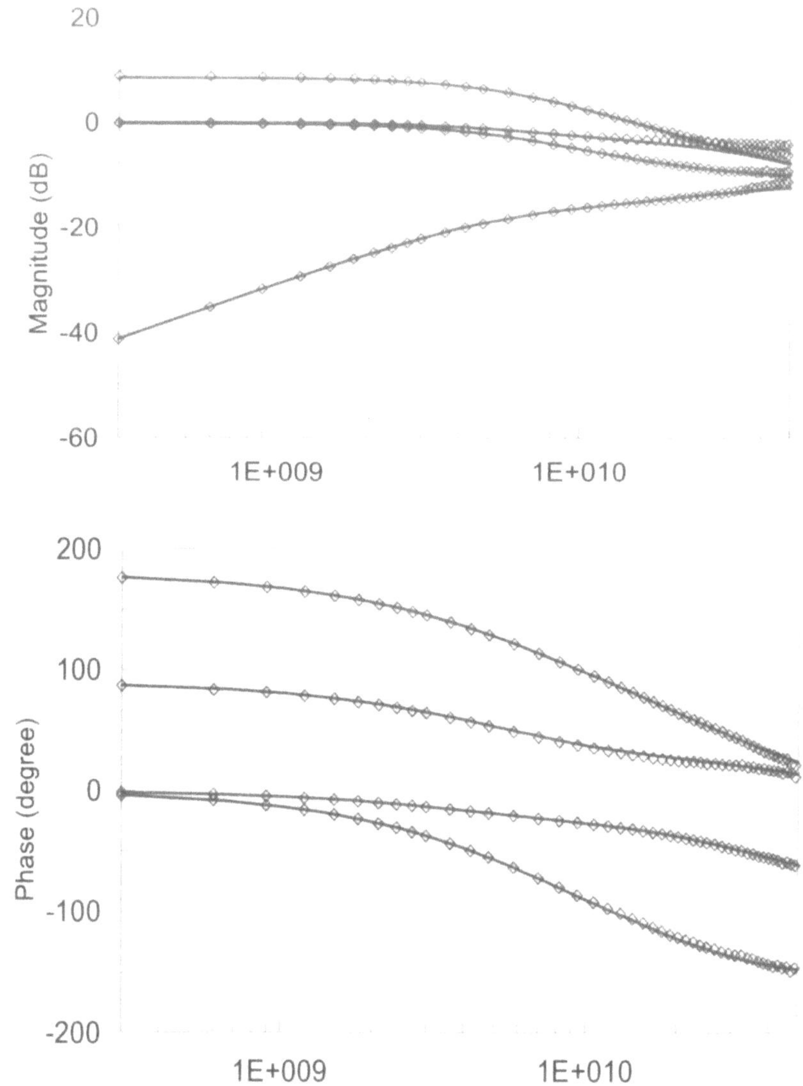

Figure 28. Comparison between simulated and measured S parameters at I_c=0.8 mA for the tabular model

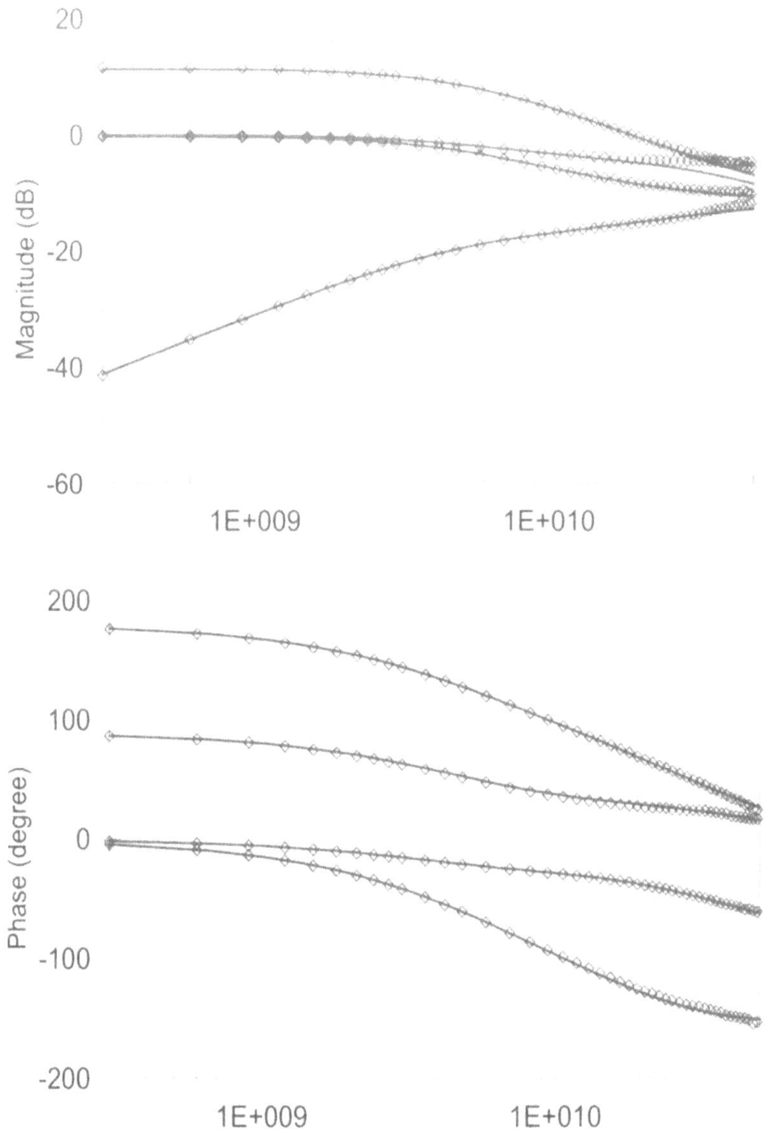

Figure 29. Comparison between simulated and measured S parameters at I_c=1.1 mA for the tabular model

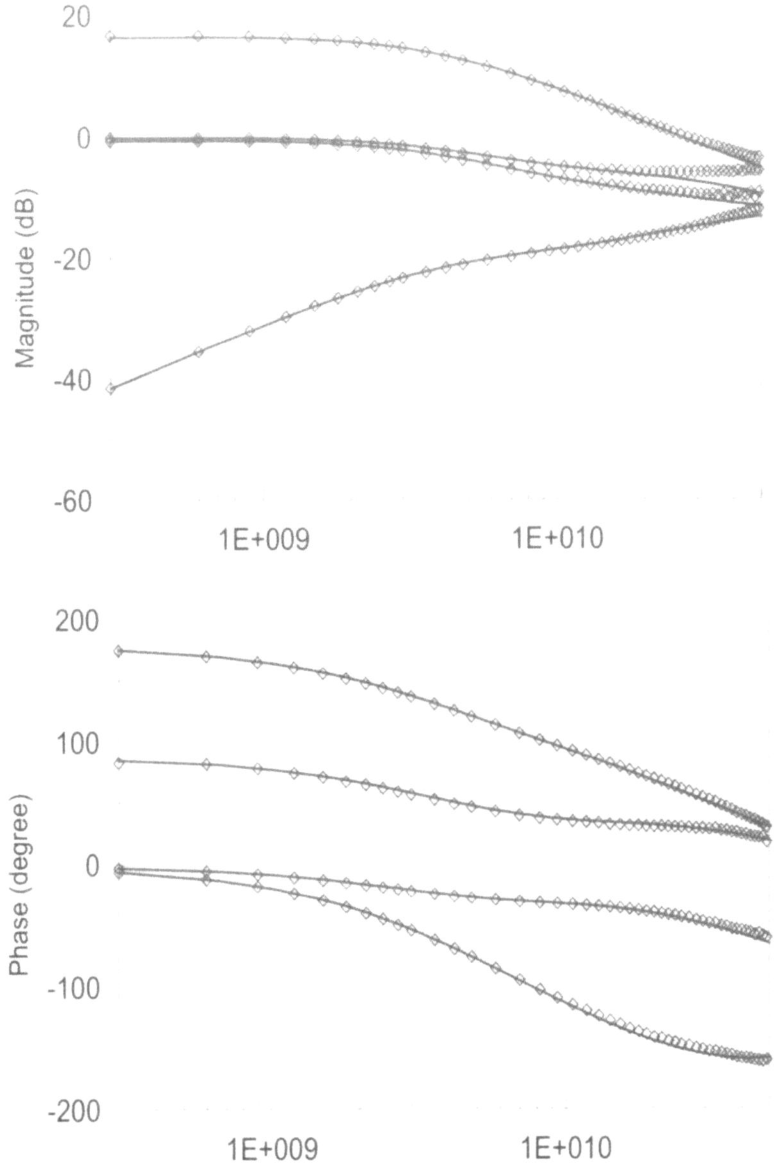

Figure 30. Comparison between simulated and measured S parameters at I_c=2.1 mA for the tabular model

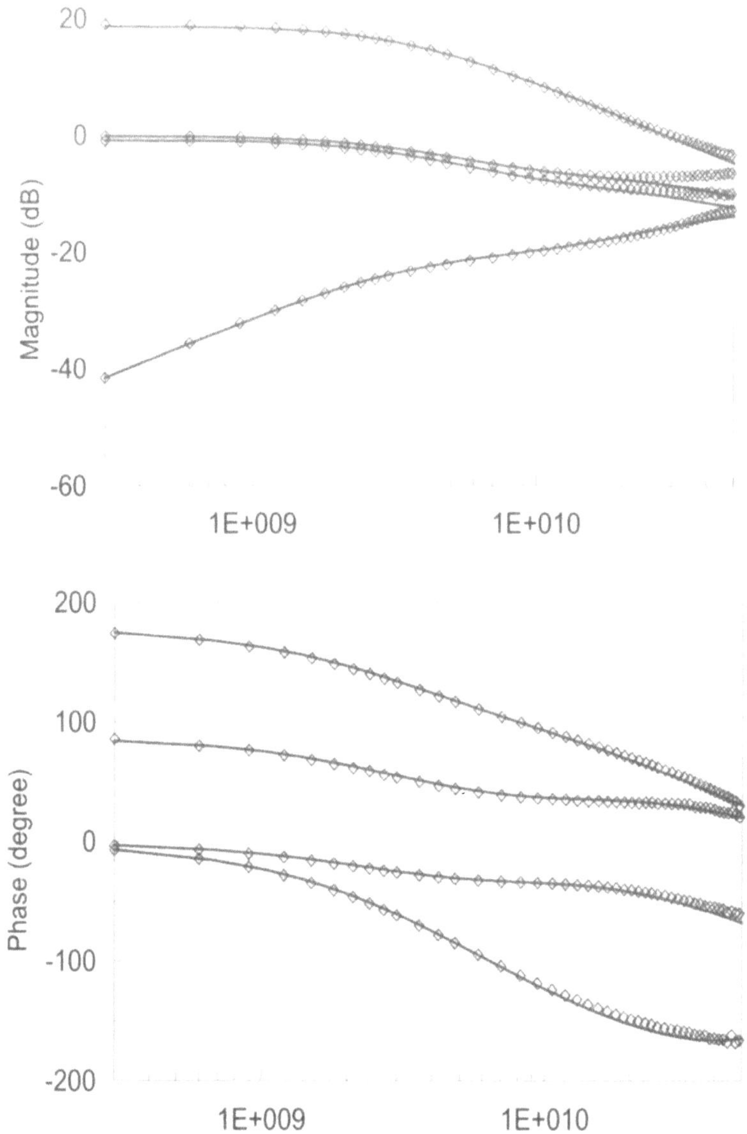

Figure 31. Comparison between simulated and measured S parameters at I_c=3.0 mA for the tabular model

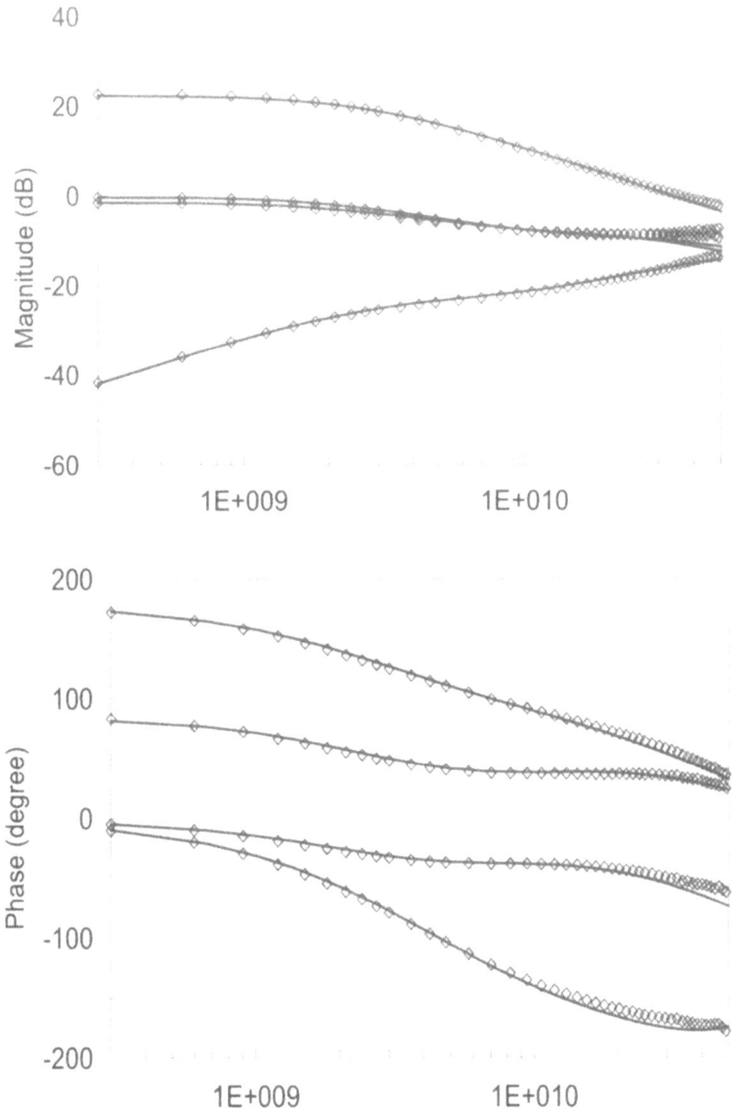

Figure 32. Comparison between simulated and measured S parameters at I_c=5.4 mA for the tabular model

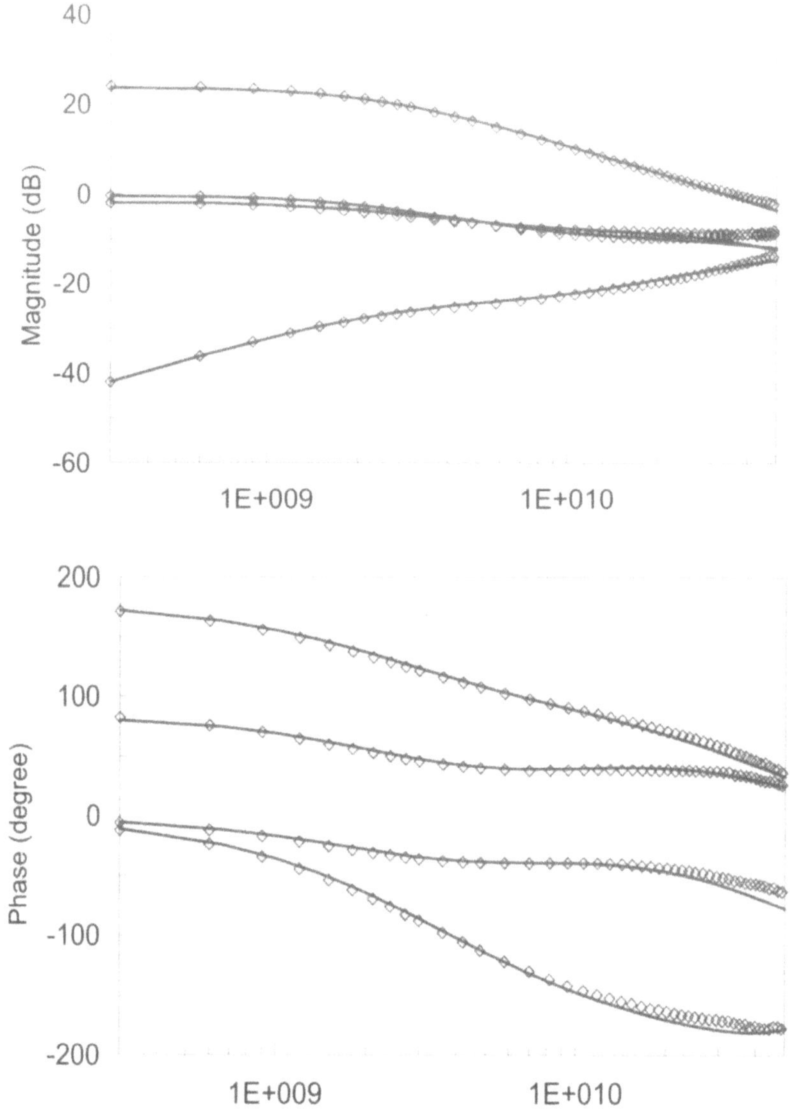

Figure 33. Comparison between simulated and measured S parameters at I_c=7.1 mA for the tabular model

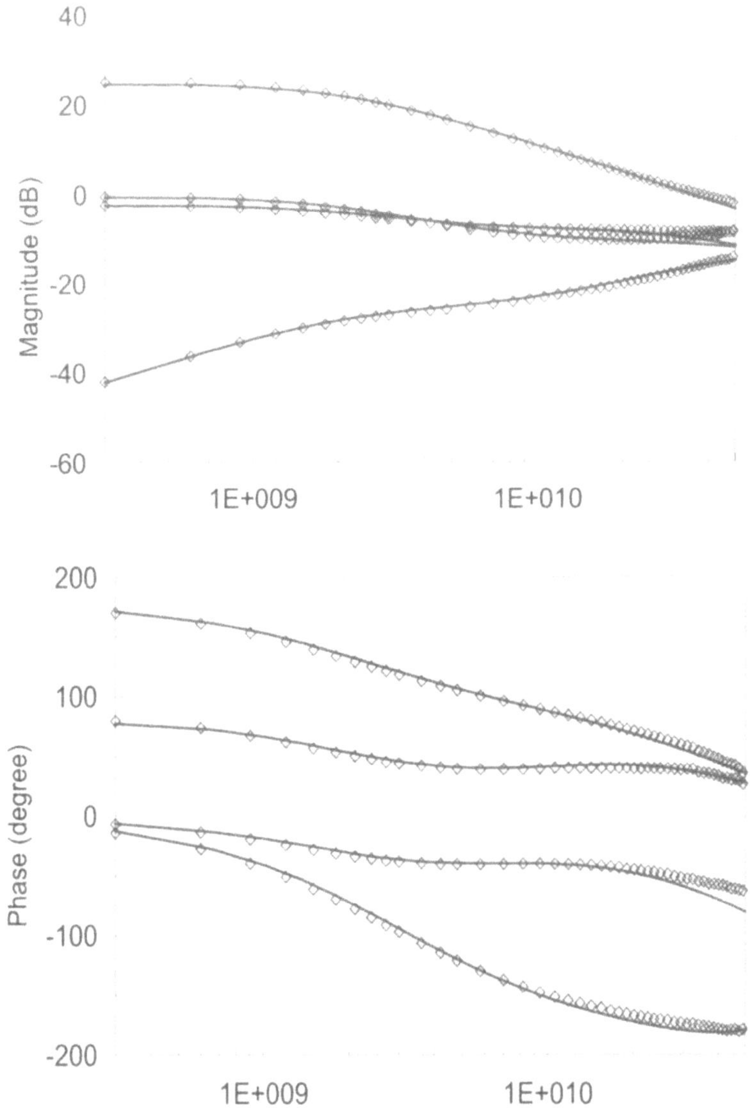

Figure 34. Comparison between simulated and measured S parameters at I_c=9.1 mA for the tabular model

In the measured bias points outside the linear region, more precisely in the high injection region, the tabular model doesn't provide useful results as it can be expected.

The parameters of that model show strong frequency dependence and it's very difficult to find a frequency range over which they behave in a nearly frequency independent way.

The mismatch between simulated and measured S parameters become more and more significant as the collector current increases going toward the deep high injection region.

Macromodel

The implementation of the macromodel is a bit more involved. Even if the intrinsic part of the model is the same as the ST SPICE Gummel-Poon, all the equations have been implemented once again to create a workbench for a next compact large signal model.

As has been said before the greatest problem involved with a compact model for bipolar transistor is the loop caused by the presence of the base, emitter and collector resistances.

Actually this problem is unsolved so the implementation of that model has been made with a macromodel, whose parts have been Modelled using the C/C++ interface of HDL-A like for the tabular model.

Formulas and algorithms for the intrinsic part of the model, which is an ST SPICE Gummel-Poon model without the series resistances and the external base-collector capacitance, have been written in C++. Even if this part doesn't include any series resistances a loop has to be solved just the same. It's caused by the base pushout parameters.

The B factor that modifies the base charge q_b depends on the collector current I_C (through the normalized collector current I_{CN}) and on the base-collector voltage V_{bc}, as is shown in Equation 23.

Even if the dependence of this factor from the base-collector voltage doesn't cause any problem, the dependence from the normalized collector current leads to a loop, which has to be solved in order to provide the correct value of the collector current.

While at low current levels the effects of the base pushout factor on the collector current could be neglected, their contribute is significant in the high injection region. Being this loop caused by model parameters it's impossible to make a further split of the model in order to let the simulator solve the loop as it has been done for the loop caused by the series resistances of the model.

Nevertheless, the dependence of collector current on base pushout parameters is weaker than that caused by the series resistances, so it's possible to solve the loop even using a simple algorithm.

In order to provide a quicker convergence and to make more compact the model, this algorithm has been implemented in C++ instead of using the internal solver provided by HDL-A.

This algorithm has been validated comparing the measured and simulated DC currents with parameters extracted by Gummel plot fitting. These comparisons are shown in Figures 35 and 36.

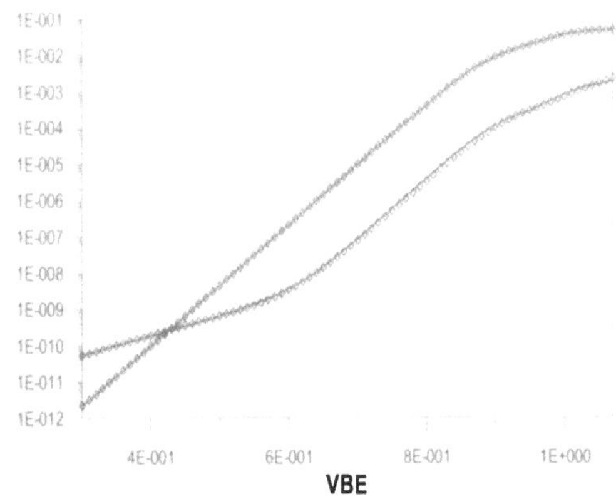

Figure 35. Comparison between simulated and measured forward Gummel plot

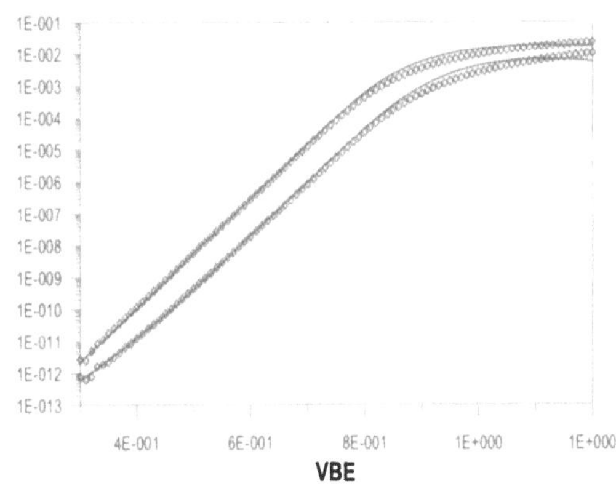

Figure 36. Comparison between simulated and measured reverse Gummel plot

The loop caused by the series resistances of the model can't be solved in the same way, in fact the presence of these resistances causes stronger variations of both the base and collector currents and thus that simple algorithm doesn't apply in this case.

In fact even if at low current levels it allows to find a solution with few iterations, the same isn't true at higher current levels when the voltage drop across the series resistances becomes comparable with the external voltage across the base-emitter or the base-collector junction.

In this case a stable solution can't be found even after a large number of iterations. For this reason the model has to be split as explained in the section dedicated to the description of the macromodel and shown in Figure 15, separating the series resistances by the rest of the components and letting the simulator solve the loop.

While the tabular model used a database file to interpolate small signal parameters, the macromodel uses a SPICE-like MODEL card in which are stored the values of parameters that have been extracted by curve fitting. A C++ routine has been included in the source code to read the model parameter file and to assign their values to static variables avoiding the loss of information between two consecutive calls of the C++ functions.

The HDL-A source code for that model is divided into two parts (DC behaviour and AC behaviour) like the one for the tabular model.

During the DC analysis the base, emitter and collector currents are calculated iterating the algorithm described above and using Equations 24, 25 and 26.

Then small signal parameters are calculated using Equations 9, 10, 11 and 12 and from 31 to 41. The values of small signal parameters are stored in static variables to maintain their value when the AC analysis will be run.

A comparison between measured and simulated S parameters in eight bias points within the forward active region of the device are shown in Figures 37 to 44.

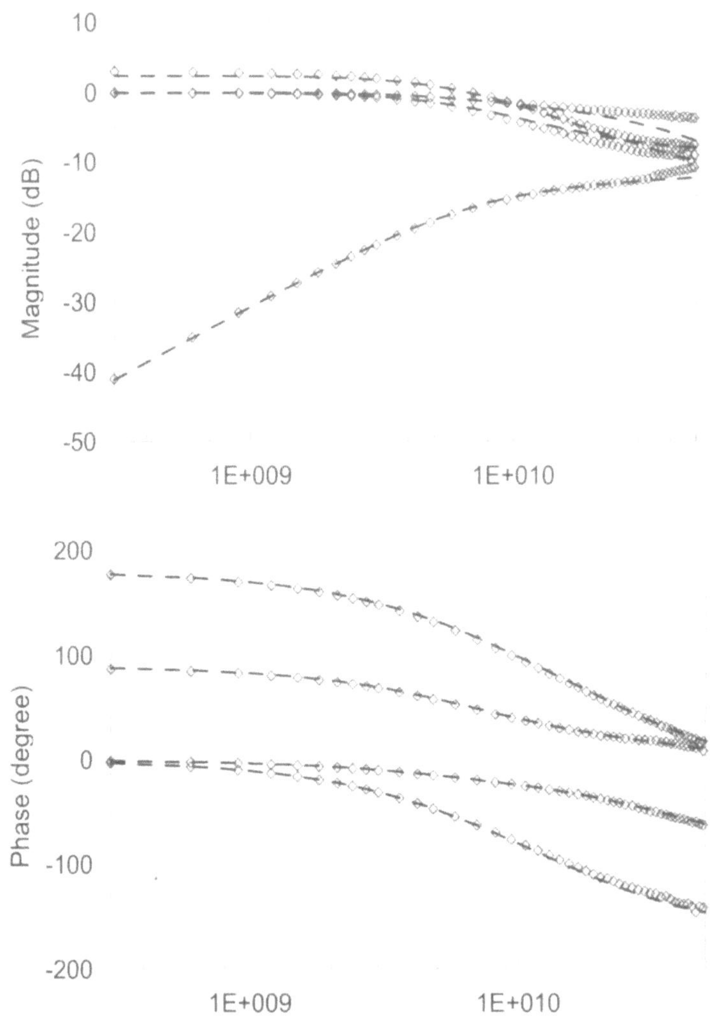

Figure 37. Comparison between simulated and measured S parameters at I_c=0.4 mA for the macromodel

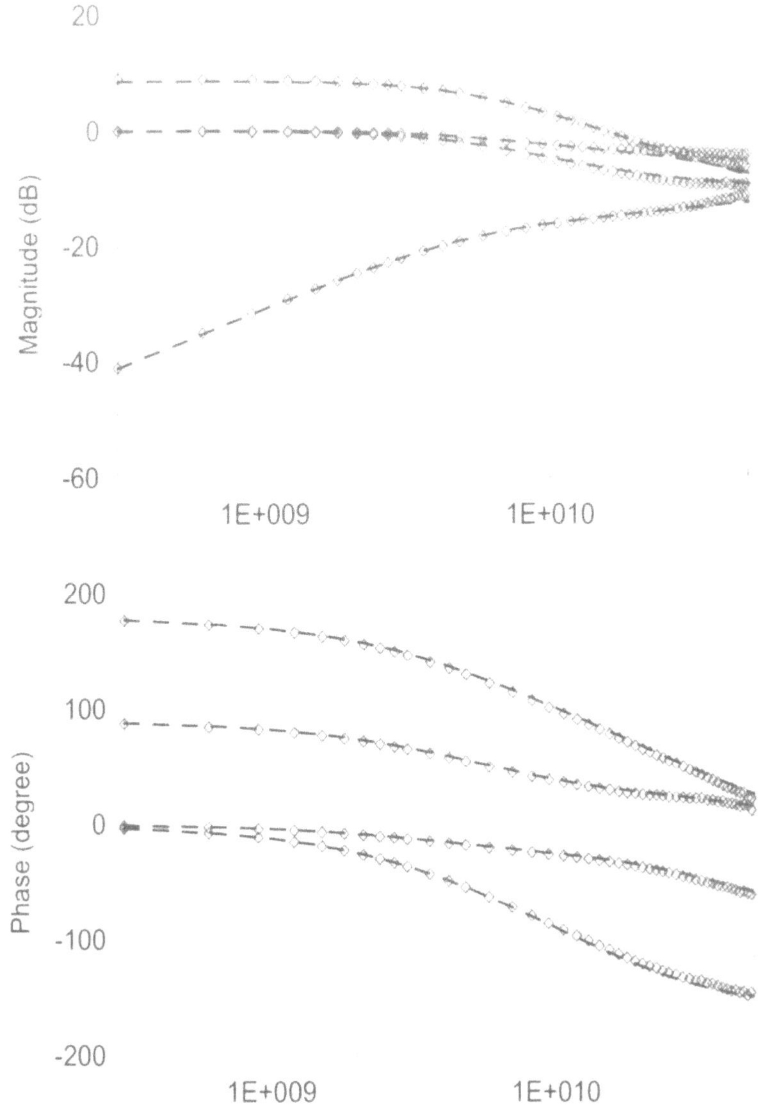

Figure 38. Comparison between simulated and measured S parameters at I_c=0.8 mA for the macromodel

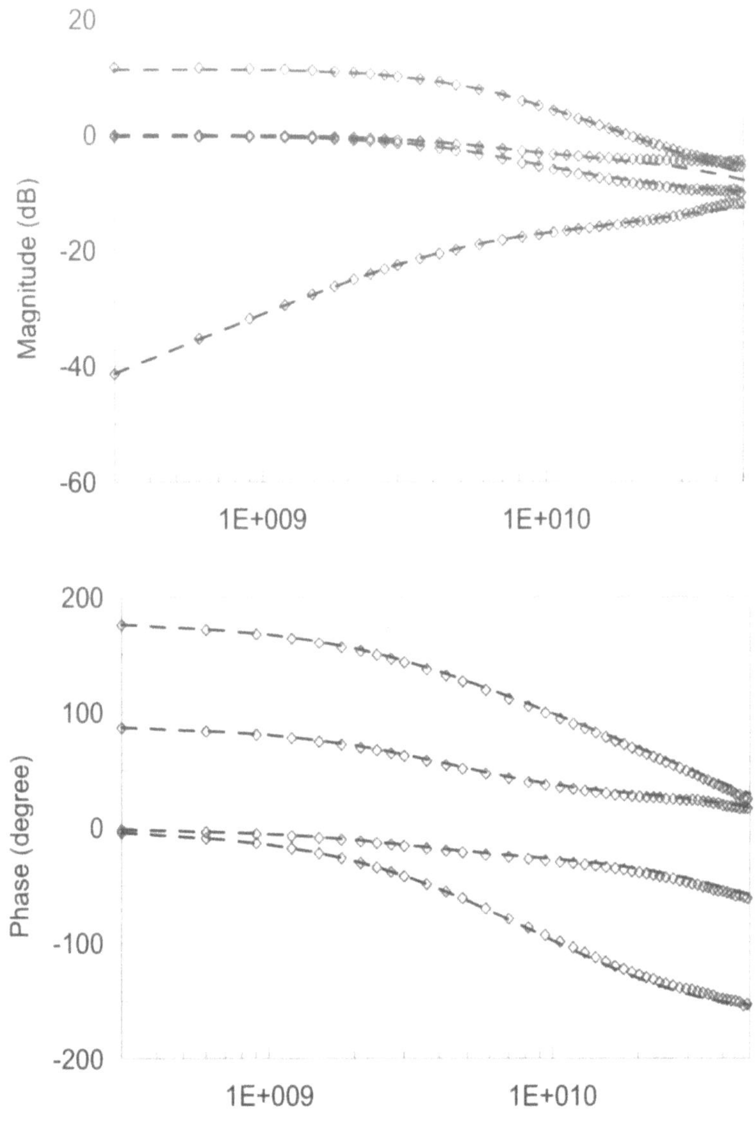

Figure 39. Comparison between simulated and measured S parameters at I_c=1.1 mA for the macromodel

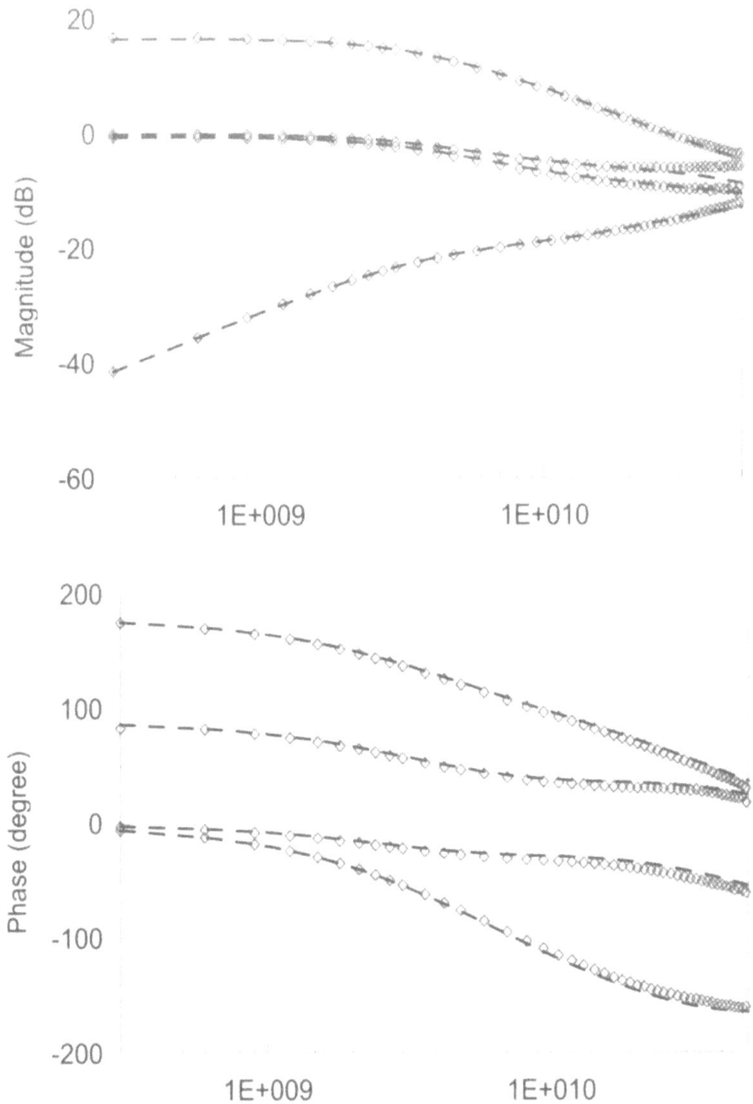

Figure 40. Comparison between simulated and measured S parameters at I_c=2.1 mA for the macromodel

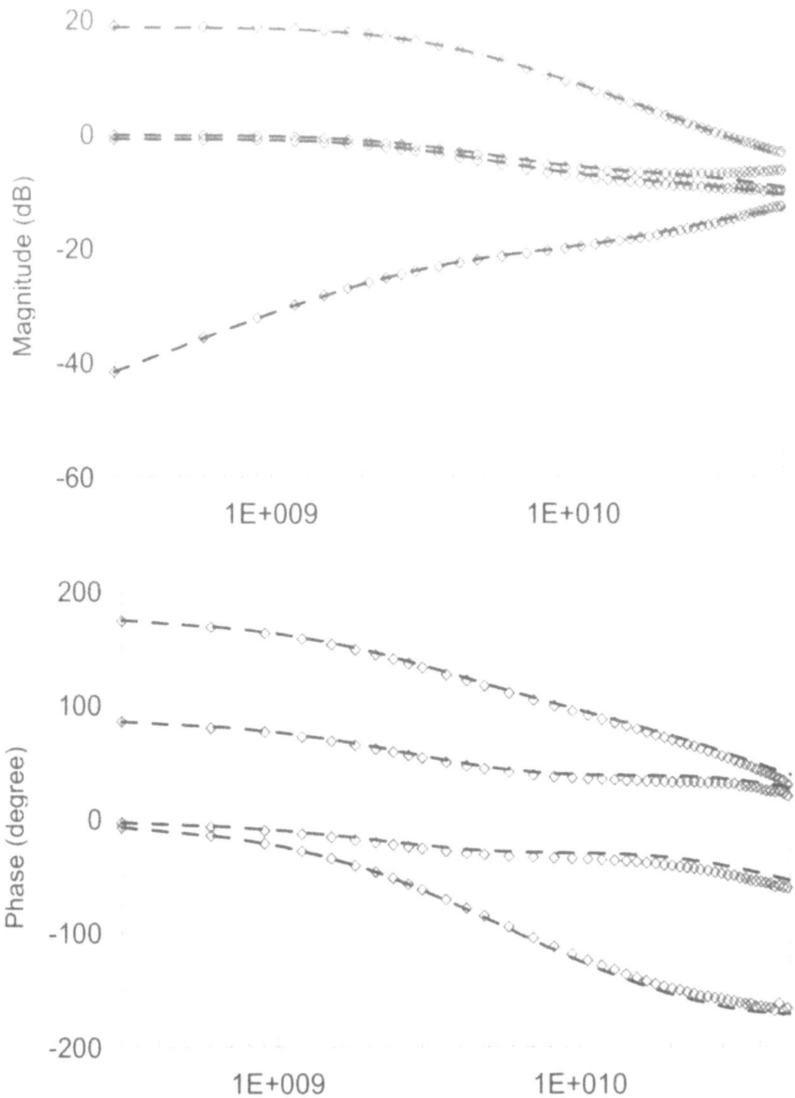

Figure 41. Comparison between simulated and measured S parameters at I_c=3.0 mA for the macromodel

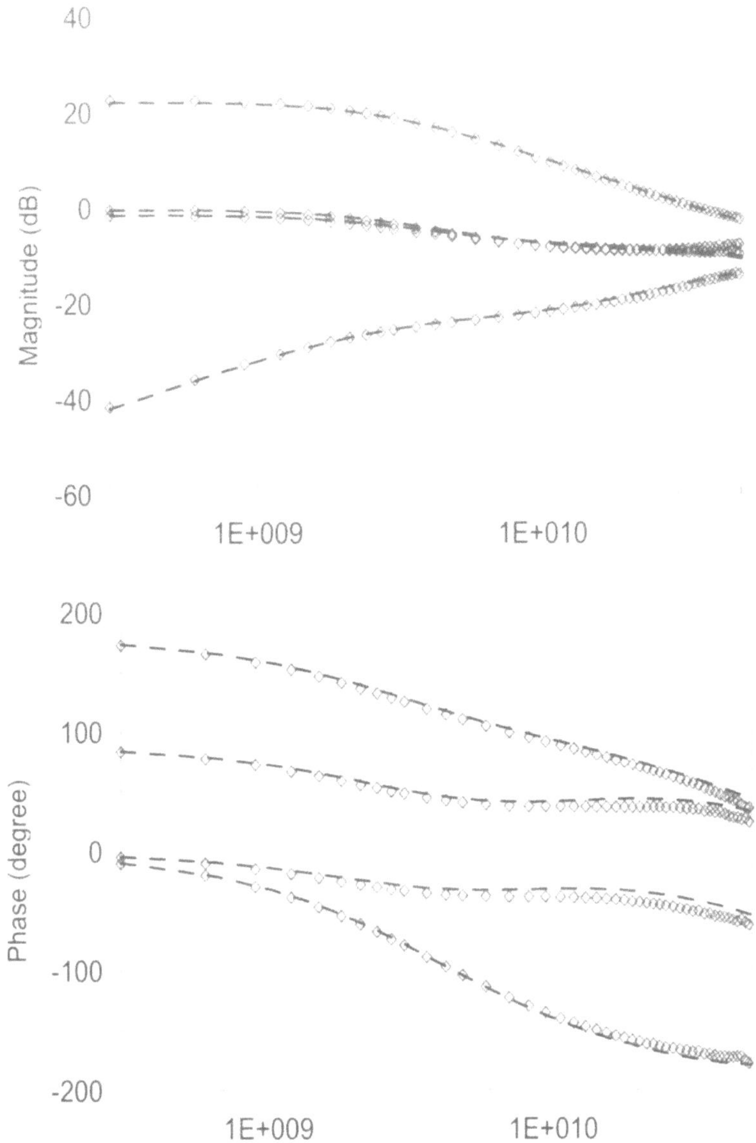

Figure 42. Comparison between simulated and measured S parameters at I_c=5.4 mA for the macromodel

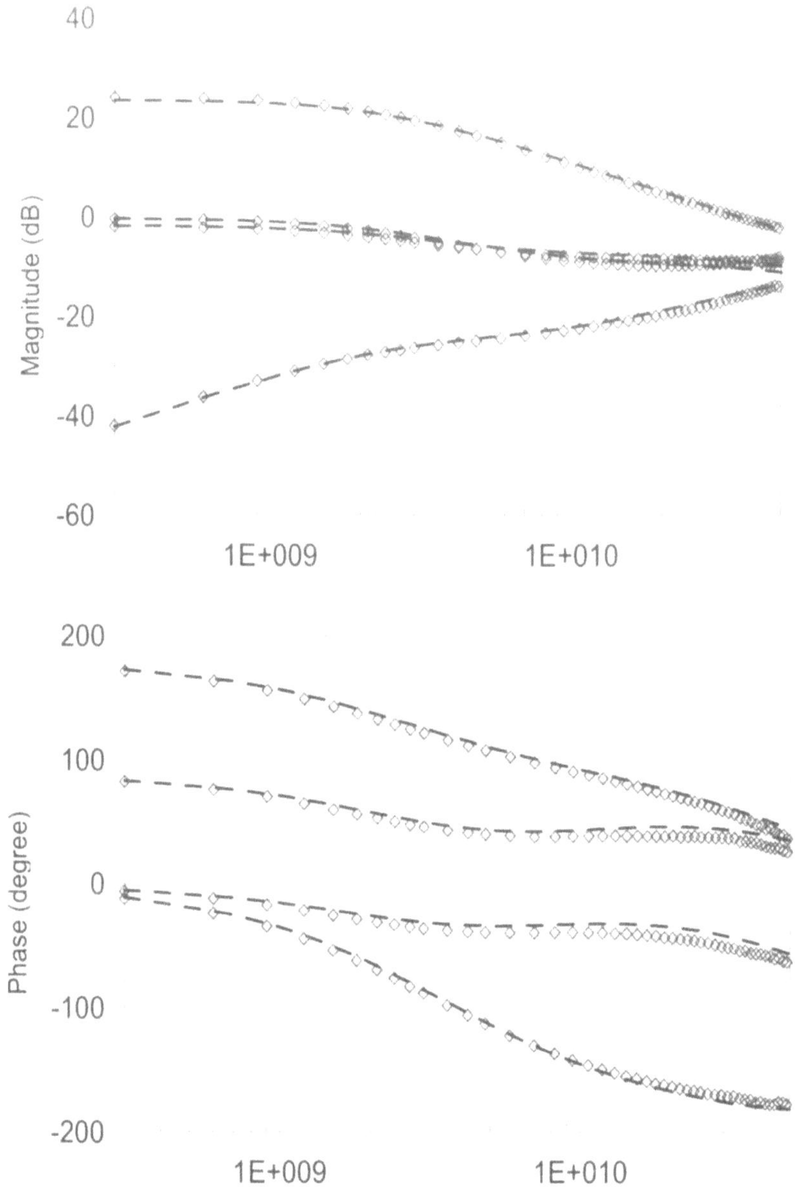

Figure 43. Comparison between simulated and measured S parameters at I_c=7.1 mA for the macromodel

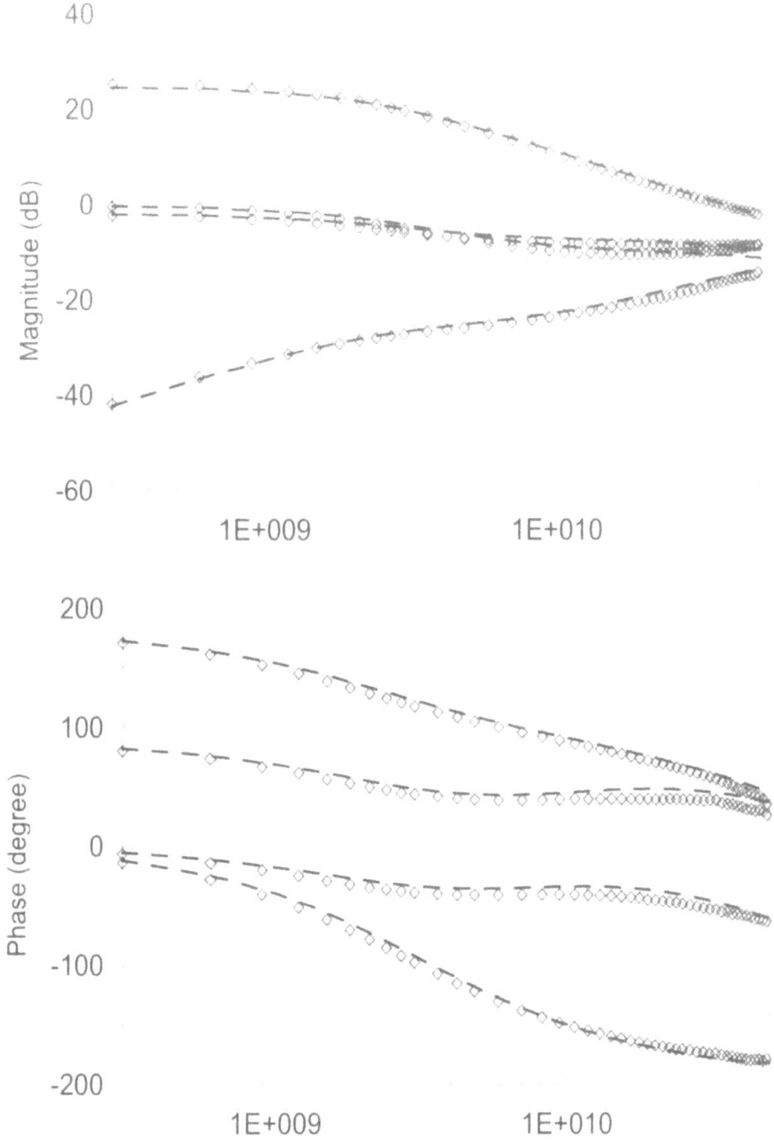

Figure 44. Comparison between simulated and measured S parameters at I_c=9.1 mA for the macromodel

It's clear that this model allows obtaining results comparable to that of the tabular model, over the entire frequency range, if the device is working in the forward active region. This is true even if the small signal parameters of this model doesn't coincide with the extracted ones owing to inevitable errors due to curve fitting, as shown in Figures 16 to 26.

When bias points outside the forward active region are to be simulated the performances of that model are quite poor, but it couldn't be in a different way. In fact in these bias points the performances of the extracted model itself can't be accepted so the model approximating the extracted one can't perform in a better way.

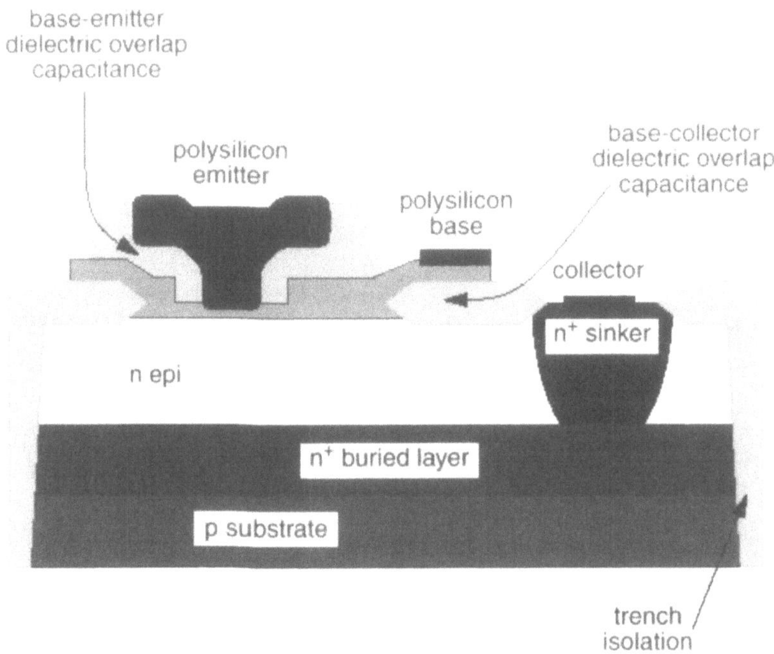

Figure 45. Vertical section of high frequency bipolar transistor technology

The cause of that mismatch is obviously the strong frequency dependence of extracted model parameters, which leads to think that the topology of the small signal model has to be changed in order to provide better performances.

An improvement to the topology could be made introducing the pnp parasitic bipolar transistor taking into account the charge injection coming from the substrate when the device is going to work in quasi-saturation region, when the n^+-p collector-substrate junction shown in Figure 45 is weakly reverse biased.

Also the dynamic emitter crowding effect have to be Modelled splitting the base-emitter capacitance into two parts: one between the intrinsic base and the intrinsic emitter and another between the extrinsic base and the intrinsic emitter. Indeed while the device is going toward the high injection region most of the current flows through the periphery of the emitter owing to the voltage drop across the distributed base resistance.

Thus the base-emitter capacitance, which in the real structure is distributed across the emitter periphery, works only in that region. Improvements to the macro-model in the high injection region and the extension to the dynamic large signal behaviour are still subject of study.

Another problem on which has to be fully thrown light is the de-embedding procedure. Owing to the ground-signal-ground structure used to make possible the measurement on the device, which is shown in Figure 46, parasitics effects influences the measured scattering parameters.

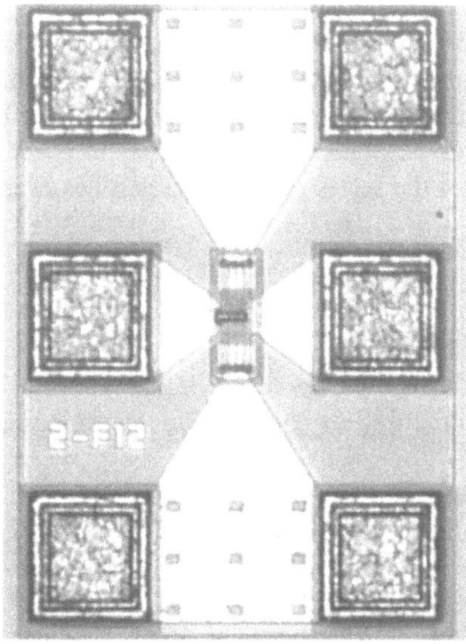

Figure 46. Microphotography of the integrated structure for RF measurement

Thus they have to be quantified and subtracted from unde-embedded data in order to provide the network parameters of the device alone. Many papers have been written in literature trying to solve this problem but it seems to be still open.

Conclusions

The starting point of this work has been the parameter extraction technique implemented by S. Rinaudo in STMicroelectronics, Catania site. This technique allows the determination of parameters of the equivalent small signal model for the bipolar transistor in the forward active region providing simulated network parameters to be quite the same as measured ones.

This technique has been extensively exploited to create multibias models in which bias dependence of parameters have a continuous behaviour in order to be successfully used in every bias point with the forward active region.

Two different models, the tabular model and the macromodel, have been presented both of them showing very good performances the AC analysis over a frequency range exceeding the transition frequency of the device being Modelled.

These models have been implemented in a SPICE-like simulator showing performances comparable to that of the extracted model over a wide range of bias levels. The noise performances of that model couldn't have been tested because the tool (HDL-A) used to implement them in ELDO doesn't allow the noise analysis to be performed. Nevertheless it's reasonable to think that the noise performances of this models will be similar to that of the extracted model they approximate, which are reported in [1].

The extension of the macromodel to the dynamic large signal behaviour is still a subject of study, in fact the high frequency behaviour of the device in the high injection region lead to problems that haven't been yet solved. The same is to be said about the implementation of a compact model, which have to substitute the macromodel, in this case a robust algorithm is essential to avoid convergence problems caused by the presence of the series resistances, as it has been explained before.

References

[1] S. Rinaudo, G. Privitera, G. Ferla, A. Galluzzo. Small-Signal and Noise Modelling of Submicrometer Self Aligned Bipolar Transistor. IEEE RAWCON'98 Proceedings, 1998.

[2] S. Lee, S. W. Kang. A Parameter Extraction Method using Cutoff Measurement for a Large-Scale HSPICE Model of HBT's. IEEE Transaction on Electron Devices, 41 (1), January 1994.

[3] S. Lee, A. Gopinath. Parameter Extraction Technique for HBT Equivalent Circuit Using Cutoff Mode Measurement. IEEE Transaction on Microwave Theory and Techniques, 40 (3), March 1992.

[4] S. Lee. Fast and Efficient Extraction of HBT Model Parameters Using Multibias S-Parameter Sets. IEEE Transaction on Microwave Theory and Techniques, 44 (8), August 1996.

[5] S. Lee, A. Gopinath. New Circuit Model for RF Probe Pads and Interconnection of HBT Equivalent Circuits. IEEE Electron Device Letters, 12 (10), October 1991.

[6] A. Kameyama, A. Massengale, C. Dai, J. S. Harris. Analysis of Device Parameters for pnp-Type AlGaAs/GaAs HBT's Including High-Injection Using New Direct Parameter Extraction. IEEE Transaction on Electron Devices, 44 (1), January 1997.

[7] B. Li, S. Prasad, L. W. Yang, S. C. Wang. A Semianalytical Parameter-Extraction Procedure for HBT Equivalent Circuit. IEEE Transaction on Microwave Theory and Techniques, 46 (10), October 1998.

[8] B. Li, S. Prasad. Basic Expressions and Approximations in Small-Signal Parameter Extraction for HBT's. IEEE Transaction on Microwave Theory and Techniques, 47 (5), May 1999.

[9] U. Schaper, B. Holzapfl. Analytical Parameter Extraction of the HBT Equivalent Circuit with T-Like Topology from Measured S-Parameters. IEEE Transaction on Microwave Theory and Techniques, 43 (3), March 1995.

[10] S. Lee, B. Ryum, S. W. Kang. A New Parameters Extraction Technique for Small-Signal Equivalent Circuit of Polysilicon Emitter Bipolar Transistors. IEEE Transaction on Electron Devices, 41 (2), February 1994.

[11] D. Costa, W. U. Liu, J. S. Harris. Direct Extraction of the AlGaAs/GaAs Heterojunction Bipolar Transistor Small-Signal Equivalent Circuit. IEEE Transaction on Electron Devices, 38 (9), September 1991.

[12] D. R. Pehlke, D. Pavlidis. Evaluation of the Factor Determining HBT High-Frequency Performance by Direct Analysis of S-Parameter Data. IEEE Transaction on Microwave Theory and Techniques, 40 (12), December 1992.

[13] D. Celi, "BJT Model Equations V2.0", UNICAD, September 1991.
[14] E. Mazaleyrat, D. Celi, A. Juge, B. Cialdella. A New Bipolar Extraction Tool for Wide Range of Device Behaviours. Proc. IEEE 1991 Int. Conference on Microelectronic Test Structure, 4, March 1991.
[15] ANACAD. Eldo Device Equation Manual. Anacad, 1999.

Design of an Integrated Optocoupler Based on Silicon Technology

Giuseppe Longo

STMicroelectronics S.r.l.
Stradale Primosole 50, 95100 Catania – Italy

e-mail: peppe.longo@st.com

Introduction

This work is aimed to design a prototype for the monolithic integration in Si of an Optocoupler (Optoisolator) device to be used in several applications such as data transmission from master to slave in control systems or for communication between power device and control logic.

A schematic picture of the whole device is shown in Figure 1. It consists of two Si areas separated by a thick oxide.

The communication between these two areas occurs by photons emitted by the LED, travelling through the SiON waveguide and reaching the photodetector.

The design is proposed to integrate in the Figure 1 scheme, some devices and functions already developed, operating at a wavelength of 1.54 um; they are:

- erbium-doped Si LED;
- Schottky photodetector ;
- an insulation bulk consisting of a thick SiO_2 layer, allowing to achieve insulation up to a voltage of 2000 V

Now are added some innovative structures[1] needed to realize an optocoupler, specifically:

- SiON isolation waveguide (to connect the two electrically insulated Si areas);

---

[1] In large part of novel projecting origin. They are to submit en route for simulations and experimental measurements on proposed apparatuses to implement.

- a lateral emitting LED (*i.e.*, emitting light along the lightguide rather than from its surface);
- source light back-standing RIE Mirrors (to enhance coupling between the LED and the waveguide);
- ARROW waveguides.

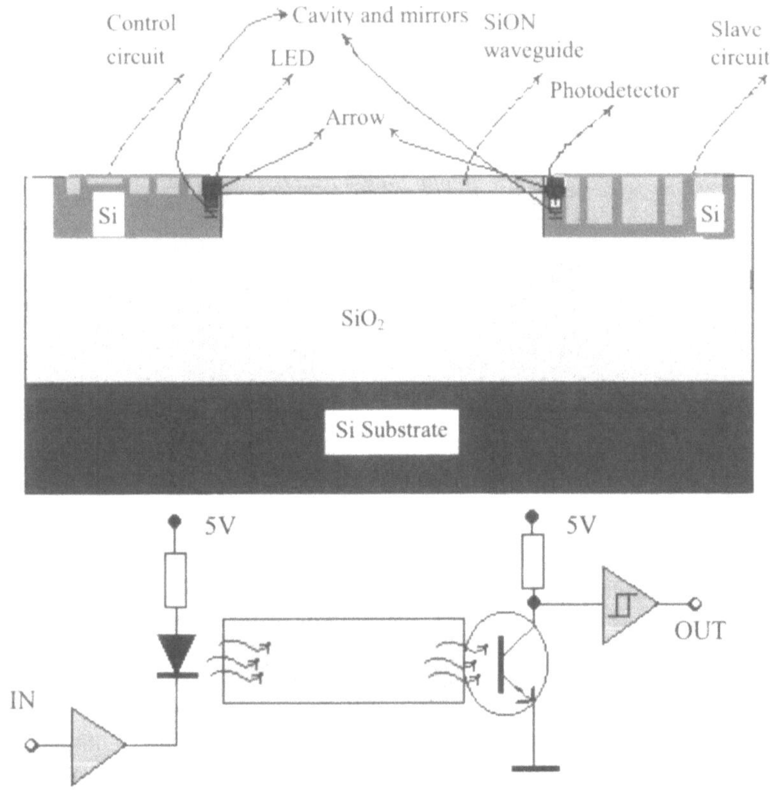

Figure 1. Optocoupler schematic and its equivalent circuit

The last feature (which also allows frequencies demultiplexing) is proposed for a more advanced phase of development (high speed - high volume data transfer applications) and will not be described in the following.

The main feature of the optocoupler is the thick SiO_2 layer; its structure is obtained by three dense series (I° II° III° in Figure 2) of deep trenches in a bulk of *n* Silicon wafer containing a buried p-type layer. Using electrochemical etch, Si is transformed in Porous Silicon.

Finally, the PorSi is oxidized to form the SiO_2 insulating bulk. The result will be two *n* Si islands in a SiO_2 sea (see the cross section below reported).

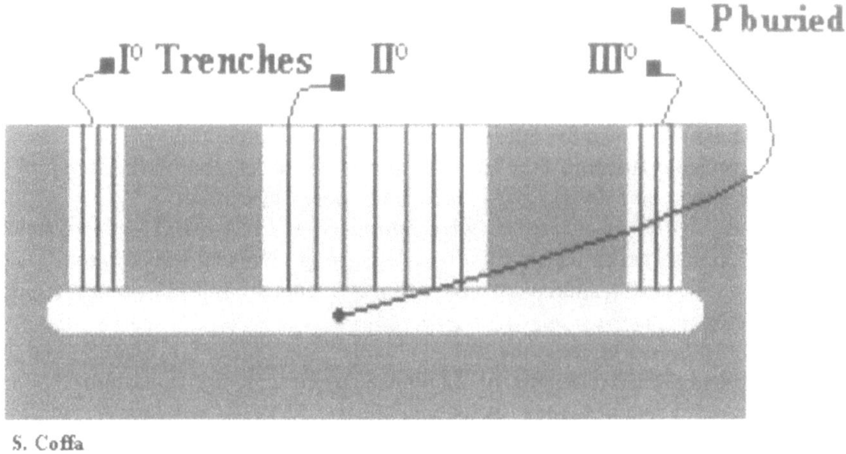

Figure 2. Insulation scheme of Optocoupler

To achieve the planar structure we need that the Er-LED should be of *Edge-Emitting* kind (EELED) versus the already existent vertical surface-emitting one. The standard LED will be replaced by a MESA diode structure laterally defined by two lateral trenches (Principal-Channel-Trenches) as is showed in Figure 6.

The design presented in this work use, as boundary conditions, the restraints related to STMicroelectronics[2] technologies, materials resources and equipment capabilities.

We developed these three major attempts:

- coupling of the LED with SiON-waveguide;
- improvement of Edge-Emitting LED progress using the RIE mirrors dimensioning;
- innovative proposal of a future alternative to the currently existing Schottky photodetector.

In addition, in paragraph III.3 a calculation related to an eventual adoption of an SiON ARC (Anti Reflecting Coating) was developed.

Properties of the Discrete Devices

The Er-doped-Si LED

To obtain light emission, we avoided using III-V semiconductors or other materials different from silicon, thanks to Er_{68} implantation at about 10^{19} Er/cm^3 in the depletion region of p^+- n^+ Si diode (which in intrinsic form is not an efficient radiation source) [5], [6].

---

[2] Site of Catania

Erbium in Si emission mechanism is based on 1.54 μm electronic spontaneous down-transition from the first excited state ${}^4I_{13/2}$ (belonging in a degenerate level[3] of Er^{3+} ions[4] incorporated in silicon), to the ground state ${}^4I_{15/2}$ level. All this is possible because Er^{3+} is now interacting with the Si lattice and will have mixed states with opposite parity; in fact this emission is not observable in the free ions in which all the states have equal parity (there is no more Stark effect).

The the Er:Si LED performance were optimised at CNR-IMETEM in Catania, using the codoping mechanism with VI or VII group elements like oxygen (O) or fluorine (F) at about 10^{20} Atoms/cm^3. Also this elements are introduced in the diode depletion region.

The O (F) codoping benefits the spread chemical coexistence of Er in Si, and by reason of their high electronegativity, facilitates the Er ionizing. In definitive halide codoping improves luminescence intensity [1], [2].

This solution permitted us to achieve acceptable room temperature efficiency and an reasonable electrical frequency modulation (up to few KiloHertz).

Moreover it was observed that the LED electroluminescence is ten times higher under high reverse bias than under forward bias; as a matter of fact, with high reverse polarization we obtain appropriate, and in great number, Er excitations by hot electrons collision, with an very small number of unradiative downtransitions.

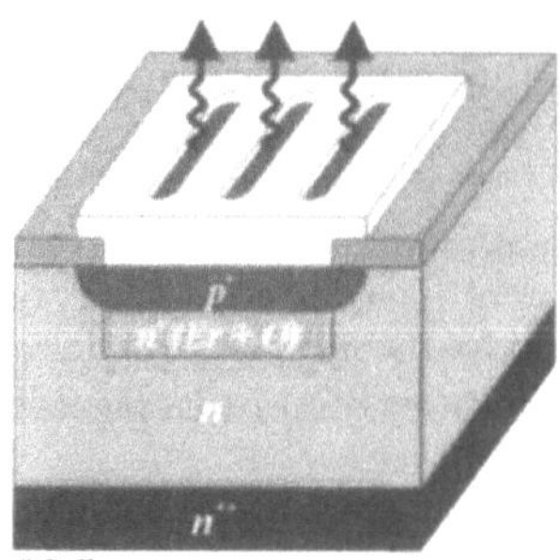

Figure 3. The first vertical-emitting Si-Er^{3+} LED

In order to use this LED in the Optocoupler several improvements need to be achieved. First of all we have to obtain lateral emission along longitudinal axis of the lightguide. Moreover the potential causes of light absorption, such as the top contact, should be minimized.

---

[3] The degeneration is created by spin-spin and spin-orbit coupling in different possible electronic configurations within the incomplete $4f^{11}$ shell. The 4f shell (outwardly shielded by an fifth level octet) has many folding level structure due to electric interaction with Si crystal. The radiative level lifetime is about 10 ms, it is however strongly depending by host defects and impurities)

[4] To favor this ionic state, the Er implantation is carried out together with halide atoms (see the following text).

The Si Waveguide

So far Si lightguides have been developed at CNR-IMETEM. This waveguide has a Si core and a SiO_2 cladding: it is therefore not appropriate for our Optocoupler structure because it would not give us the requested light transmission in a high insulating environment. Nevertheless, it represents a valid starting point for our insulating light transmission system project at STMicroelectronics Catania site (see below).

Figure 4. Photos regarding the fabricated Si-waveguide array while is tested (left). On the right is the transmitted radiation.

The electrooptic supremacy of silicon stands in its transparency at the communications wavelengths of 1.3 μm and 1.54 μm and in its high refractive index functional to confine the radiation: in this fashion, optical circuitry can be monolithically integrated with electronic circuitry, making it suitable for optoelectronic applications at these wavelengths.

Optical separation from the substrate can be acquired mainly by two methods, restraint in poor doped epitaxy on a rich doped substrate or with the use of silicon-on-insulator (SOI) technology, which is finding growing apply in the electronics industry for digital circuits. The latter is surely preferable for lower losses and praticity.

Silicon lightguide fabrication is not too difficult; we start with straight strip of thermal 1.5 μm dioxide mask grown at 1100°C on <100> wafer. After dry etch, under the strips is formed a silicon ridge 8.5 μm tall with a top of 1.5 μm which will be the rising structure of our silicon waveguide.

To meet our target, *i.e.*, the manufacture the optocoupler, we will employ a lightguide having both insulating SiON core and SiO_2 cladding.

The Schottky Detector

A Schottky diode is formed by two metal contacts on a semiconductor of *n* doping type (metal replace so *p* region); the anodic contact is ohmic and the cathodic is rectifying. To detect a photon of frequency ν, we use its energy $h\nu$ to promote a metal electron in conduction band of Silicon from where it will be collected.

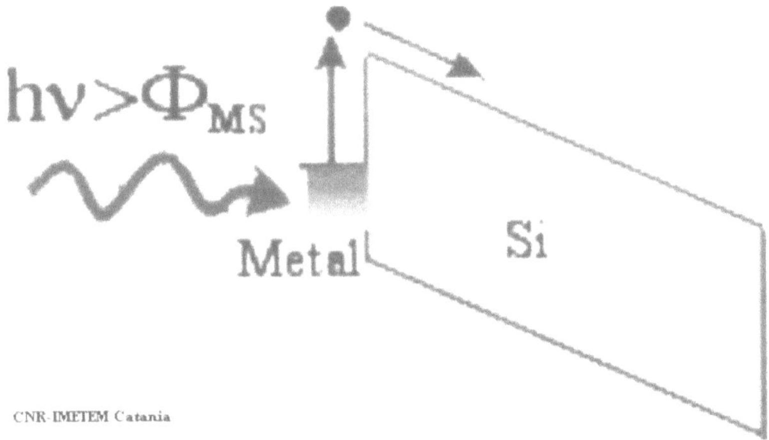

Figure 5. Picture showing the energetic band situation at metal-semiconductor interface

The barrier related to electron band promotion is

$$\Phi_{MS} = \mathbf{W} - \chi$$

then must be

$$\mathbf{W} - \chi < hc/(1.54\mu m \times 10^6) = 0.8 \text{ eV}$$

where **W** is the work function of adopted metal, χ is the electron affinity of silicon, **h** and **c** are well-known universal constants, and 1.54 μm is the erbium emitted wavelength. In last analysis, the resulting **W** is the platinum one or similar. In Figure 24 is shown our Schottky detector novel design.

Integration of the Optocoupler

A Sketch of the Structure and a Schematic Flow Chart

In this chapter will be presented the overall structure of the Optocoupler, highlighting the most innovative solution. One of these is outlined in next picture: SiON waveguide is here suggested to split in at least two branches B_1 and B_2 in order to obtain the mutal matching between both LED and detector fronts with the adopted waveguide port. In fact, LED and detector fronts are many times transversally longer (15-30 μm) than the adapted SiON lightguide port diameter (3 μm long × 2 μm of height) fitted for monomodal and low loss trasmission of radiation. A schematic of the structure is presented in Figure 6.

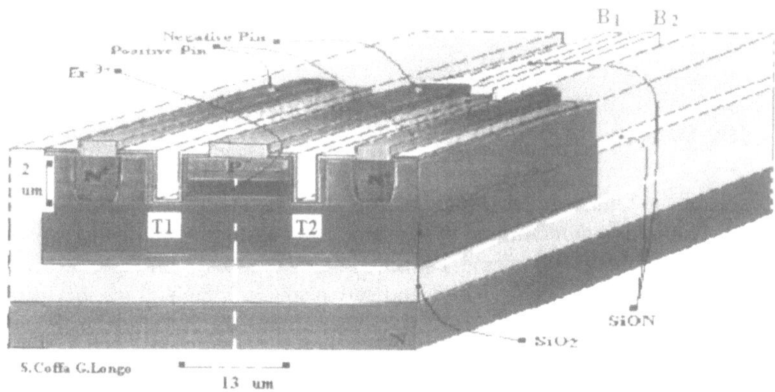

Figure 6. Schematic LED coupling with a doubled SiON-waveguide. In sight note the two Principal-Channel-Trenches T_1,T_2.

Other noteworthy features of Figure 6 are:

i. (the planar LED now consists of a MESA diode laterally defined by the T_1 and T_2 Principal-Channel-Trenches;
ii. the cathode contact is taken from the top-front of the structure.

Now we introduce a simplified scheme of the Optoisolator-associated flow chart (Schottky detector instance).

Starting Si wafer: *n-type* with a buried p-type layer

1. Insulator lithography
2. Trench etch to reach the buried p-type layer
3. Formation of porous Si
4. Porous Si oxidation
5. Anode lithography
6. Boron (B) implantation
7. Thermal process to diffuse B
8. Er implantation
9. Cathode lithography
10. Arsenic implantation to form the cathode of the diode
11. Thermal process to optically activate Er and diffuse As
12. Guide lithography
13. Etch of the SiO_2
14. SiON deposition
15. SiON lithography
16. SiON etch for guide definition
17. Contact lithography

18. Metal deposition
19. Metal lithography

The SiON Waveguide

This section of the Optocoupler (Optoisolator) must conduct the control signal photons from the source to detector and at the same time assure the galvanic isolation. A solution to this problem is offered by SiON layer on SiO_2; in fact, it offers the wanted difference of refractive index and the necessary high grade of insulation (about 10 MV/cm). A layout of a rib SiON waveguide, exhibiting low losses (less than 1 dB/cm) and single mode propagation, is shown in Figure 7.

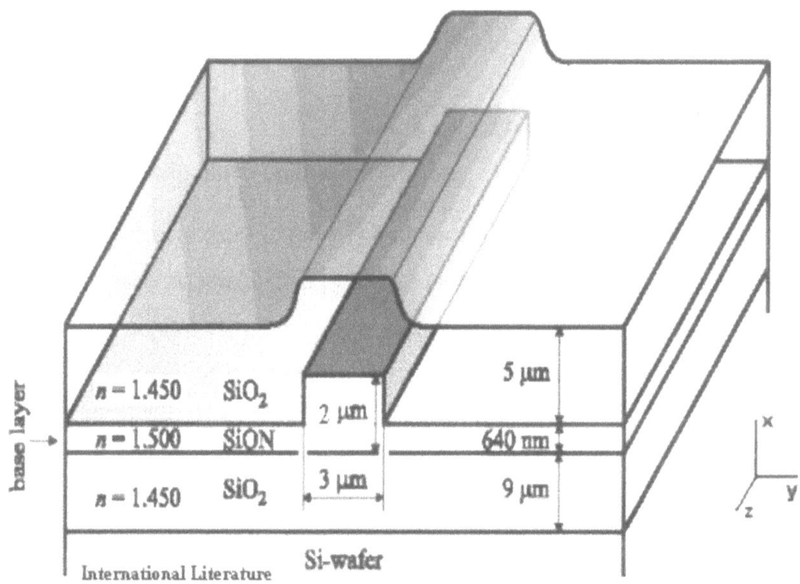

Figure 7. SiO_2-SiON rib-ridge waveguide for OEIC. With his slab basis it is optimized for monomode and low losses fashion working at 1.54 µm wavelength.

Such guiding structure is one of two subwaveguides (B_1 or B_2) outlined in Figure 6 or in Figure 15 and Figure 16 and shall be extracted by internal trenches in the strip of SiON between the Principal-Channel-Trenches.

To obtain the guided propagation the core must have refractive index n_0 higher than cladding one n_c and the angle θ that rays form with the normal to the film, must be greater than the critical angle θ_c, (total internal reflection for permitted internal modes).

Assuming a monochromatic and plane wave with angular frequency ω, free space wavelength λ, wave vector equal to $\vec{k}\, n_o$ ($k = 2\pi/\lambda = \omega/c$), and a relative index n_o ($n_o = 1.5$ while the cladding index refraction n_c is 1.45).

In first approximation, we consider a slab lightguide; the estimated results will extend to our lowered ridge waveguide. The propagation constant β is defined by the product of the free space wavevector and a new quantity called "effective guide index" N: $\beta \equiv kN \equiv kn_0 sin\theta = k_z n_0$ ($N = n_0 sin\theta$).

Not all propagation constants are allowed since there are θ angles that do not permit the radiation confinement (unguided modes). The guided modes are established by compelling the self-consistency requirement (the wave must replicates itself after each round trip along the transversal x direction to produce a constructive interference).

If h is the core average diameter, then the transverse resonance condition is

$$2kn_0 h\,cos\theta + \Phi_c(\theta) + \Phi_c(\theta) = m\pi$$

this is the sum of three phases; the first is related to one round trip on x and on y, the second and the third are concerning at the shift phase for total reflection on the top and on the bottom of cladding, m is number mode. The expression of $\Phi_c(\theta)$ is specificated by type of polarization:

$$\phi_c^{TE} = arctg\frac{-\sqrt{n_0^2 \sin^2\theta - n_0^2}}{n_0 \cos\theta} \quad \text{...for transversal electric (TE) mode}$$

$$\phi_c^{TM} = arctg(\frac{n_0^2}{n_c^2}\tan\phi_c^{TE}) \quad \text{...for transversal magnetic (TM) mode}$$

the permitted modes angles (which belong of discrete set A) arise by resonance condition when it come true with a m integer number, then from this angle θ_m we obtain the related propagation costant $\beta_m = kn_0\, sin\theta_m \equiv \beta_m(\mathrm{h},\omega)$ and $\mathrm{N} \equiv \mathrm{N_m}$. What is the range of effectives index of guided modes (and related propagation constant) ? Since $\mathrm{N_m} \equiv n_0\, sin\theta_m < n_0$ and $\mathrm{N_m} > n_c$ (for total reflection occouring), arise that $n_c < \mathrm{N_m} = n_0 sin\theta_m < n_0$.

After a simple deposition, a SiON lightguide, at 1.54 μm has a loss level equal or higher than 2 dB/cm. Nevertheless it is possible to reduce that value to less than 1 dB/cm applying a 1000°C annealing, or once more we can render this losses equal to 0.2 dB/cm with a 1150°C annealing.

This result can be explained by considering that heat treatment remove high contents of Hydrogen with its absorptions due to vibrational overtones of Si-H and N-H bonds.

We will now see why it was important for us to know what are the propagation fashions of the electromagnetic radiation in our 3 x 2 μm Rib waveguide; in fact with these informations we fixed opportune hypotheses to adopt our multibranches SiON waveguide.

For such an analysis it is usual to apply an approximation about radiation propagation in materials: Beam Propagation Method. BPM is an intuitive technique that split the transversal scattering and the longitudinal field propagation in two separate problems to obtain a realistic modes profiles. We must hypothesize that radiation moves along an optical axis (z) and transversal scattering angles take place only under a small angle of incidence with cladding (paraxial limit).

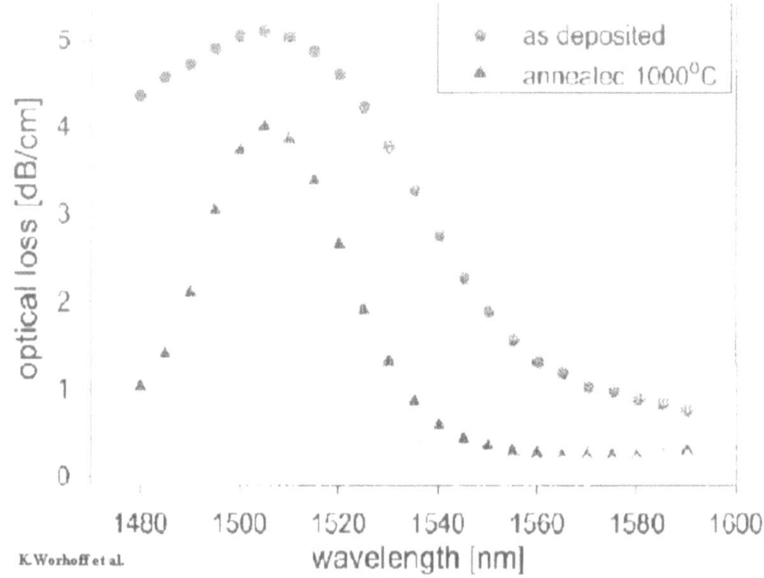

Figure 8. Optical loss of 4 μm slab SiON without and with high temperature annealing

To expose this theory briefly, we start with Maxwell equation for electric field distribution in inhomogeneus mediums (see Cartesian reference axes in Figure 7 inset):

$$\nabla^2 E(x,y,z) - n^2(x,y,z)k^2 E(x,y,z) = 0$$

for paraxial estimate we can write the following electric field eigenvalues equation:

$$j\frac{\partial E}{\partial z} = \hat{G}E \quad \text{where} \quad \hat{G} = \frac{1}{2n_0 k}\left\{\frac{\partial^2}{\partial x^2} + \frac{\partial^2}{\partial y^2} + k^2\left(n(x,y)^2 - n_0^2\right)\right\}$$

along longitudinal propagation axis *z* a constant field distribution at a unique index and with only an intensity decrease. Then (x, y) is the intensity distribution transversal plane.

Let's introduce (for a reduced numerical calculation effort) the imaginary distance ***jz*** and the reference average index n_0. Solving the last autovalues equation (field in exponential form), we can express E as linear combination of eigenfields (eigenmodes):

$$E(x,y,z) = \left(\sum_{i=1}^{\infty} c^{(i)} E^{(i)}(x,y)\right) e^{Gz} = \sum_{i=1}^{\infty} \left(c^{(i)} E^{(i)}(x,y)\right) e^{G^{(i)}z}$$

where

$$\hat{G}^{(i)} = \frac{k}{2n_0}\left(\left(n_{eig}^{(i)}\right)^2 - n_0^2\right)$$

is a "eigenβ" by which we have now, instead of index distribution n(x,y), a single value n^1_{eig} referred now to a single i-mode.

For an established analytical resolution scheme regarding autovalues equations (Crank-Nicholson), we apply the following discretization Δz along z

$$\frac{E_{n+1} - E_n}{\Delta z} = \alpha \hat{G} E_{n+1} + (1-\alpha)\hat{G} E_n$$

according to propagation steps Δz and an optimization parameter α. E_{n+1} is the unknown field at a step subsequent to one where field is known. E_0 is the field produced by LED.

Solving, from last equation we obtain

$$E_{n+1} = \sum_{i=1}^{\infty}\left[\frac{1+(1-\alpha)\hat{G}^{(i)}\Delta z}{1-\alpha\hat{G}^{(i)}\Delta z}\right]E_n^{(i)}$$

where

$$A^{(i)} = \frac{1+(1-\alpha)\hat{G}^{(i)}\Delta z}{1-\alpha\hat{G}^{(i)}\Delta z}$$

represents the amplitude factor A of i-mode.

One of its graphics is given by the following diagram in function of the product between step Δz and $G^{(i)}$.

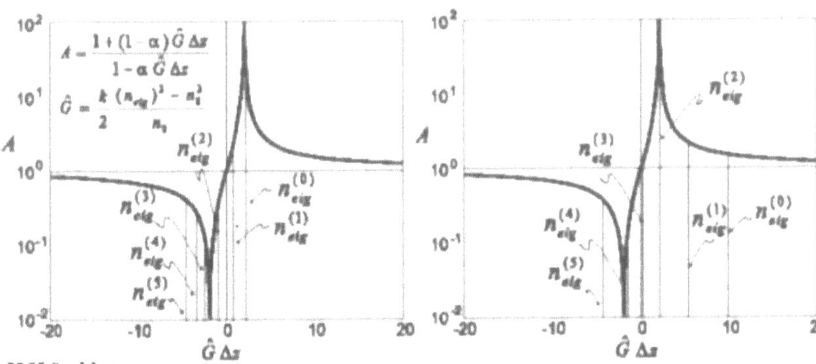

Figure 9. Amplitude factor A with a zero and a pole. By the two (left and right) sets of (G,z,α), we see what is amplification of every eigenmode with a given n^i_{eig}.

Now we can set the appropriate $\boldsymbol{n}^{i}_{eig}$ to propagate our field so that it is positioned near a pole and could enjoy of high amplitude during its propagation.

For a particular waveguide geometry, we find that more than one distribution is propagating inside the lightguide (*i.e.*, distribution orders higher than fundamental one) with two possible polarization independent states: Transversal Electric (TE) and Transversal Magnetic (they have fundamentally equivalent description faculty).

With the support of this theoretical base we can usefully refer to E = E(x,y) intensity distribution (of harmonic type) inside the waveguide to analyze the related losses. It is found that minimal losses are reached with fundamental mode propagation (zero order); it has a Gaussian shape the cross-section of which is shown in Figure 10.

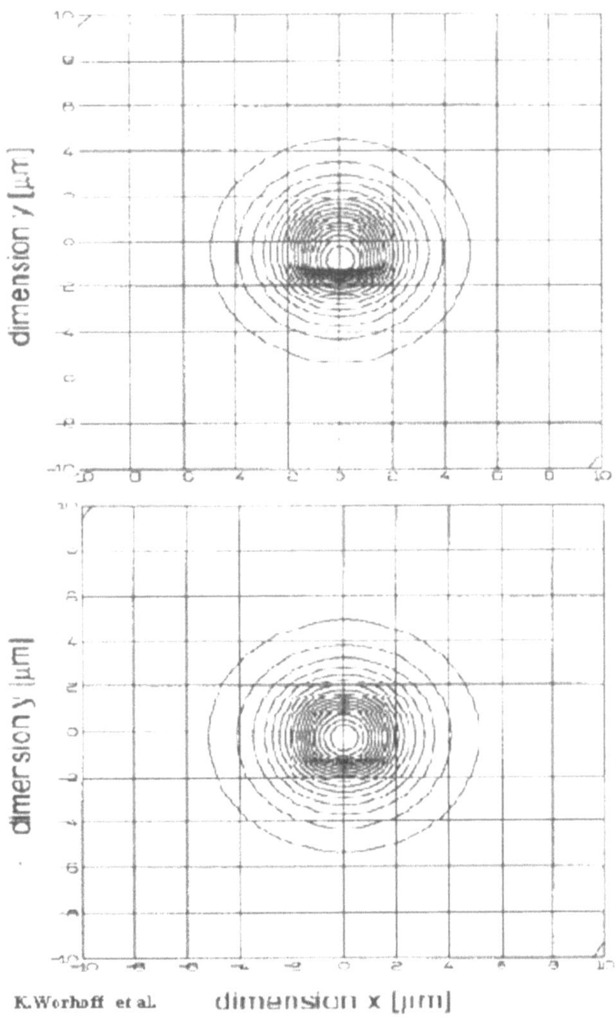

Figure 10. E(x,y) cross section distribution of fundamental mode in about 6 um^2 rectangular waveguide section for TE and TM polarization

Next we show the earliest six modes of superior order propagating simultaneously (but at reduced intensity) with the fundamental one in our ridge 3 μm x 2 μm SiON waveguide, buried in SiO_2 as superior and inferior cladding.

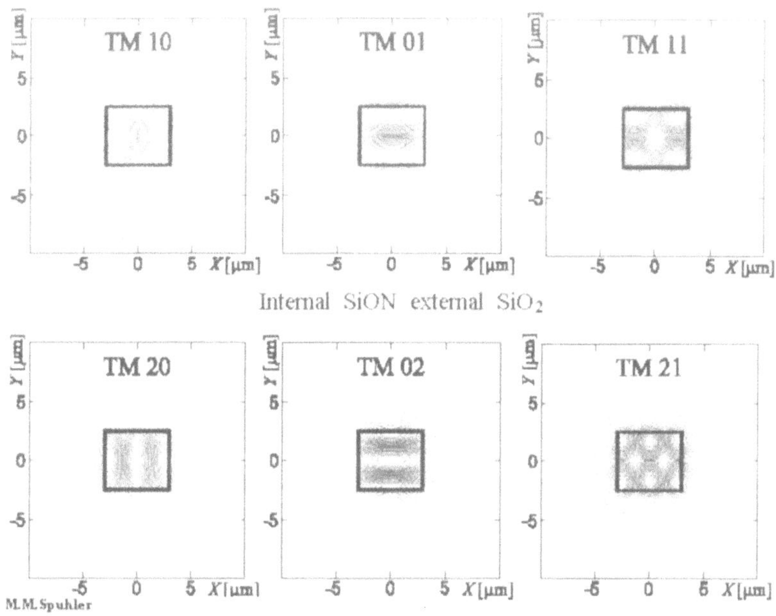

Figure 11. Six fundamental superior orders (double[5]; along x and along y) in polarization TM permitted in our isolation waveguide. The seventh order has in this transmitting structure a neglecting intensity respect to TM 00.

Note that the previous propagation modes have higher losses than the fundamental one, having more external flux lines.

Emission in the Plane and Coupling between LED and SiON Waveguide

In this section the design of the best coupling between the LED and the SiON lightguide will be presented. The work was aimed at the following targets:

- achieve only one light-emitting facet;
- achieve the maximum optical power emitted in a small solid angle and into a very narrow window.

As mentioned in the introduction, great help is given by the adoption of an Edge-Emitting source with a back reflecting DBR mirror system; we shall deal with DBR.

---

[5] The two numbers after polarization mode express the related orders respectively along x and y. The number 0 is referred to fundamental order (even) and declares that there are 0+1 maximums of field distribution, along the related axis. The number 1 is referred to first order (odd) and declares that there are 1+1 maximums of field, and so on.

Below is showed our planned light source.

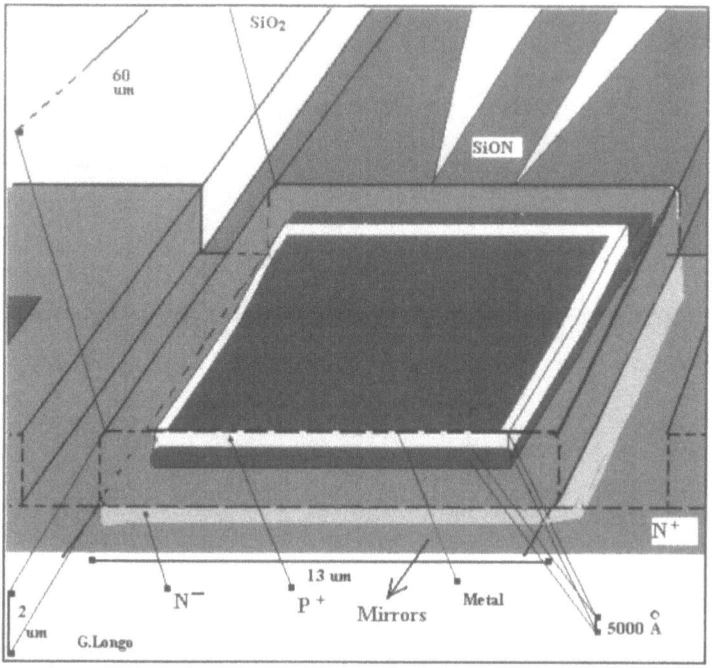

Figure 12. 3D layout of new Edge-Emitting LED. Also shown is the three-branched SiON waveguide (see farther)

Figure 12 shows a LED of about 600 um^2 of junction area; this at present is a middle-small area value for our Er LED.

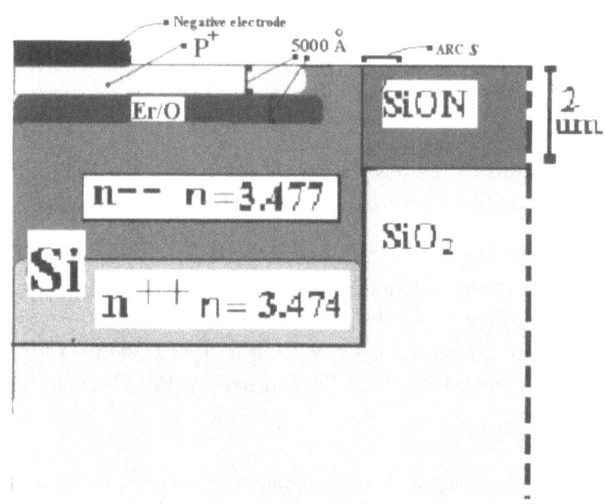

Figure 13. Si-SiON-SiO_2 boundary (cross-lateral section)

Note that with the illustrated distribution profile of Er doping and thanks to the below-standing Si n^{++} as bottom confinement layer[6], we obtain the wanted very eccentric ellipsoid radiation lobe whose horizontal major axis plane is coincident to SiON plane. This resulting ellipsoidal lobe will be further in pipe fashion laterally Modelled by the left and right Principal-Channel-Trenches. In the way it is formed the LED case-Si-lightguide that will direct the produced light through silicon to SiON.

Finally we will obtain the lobe back-limitation by using the projected DBR. The adopted model concerning our LED case-Si-waveguide, in the bottom delimited by a n^{++} layer, it is founded on Drude model illustrated in Figure 14.

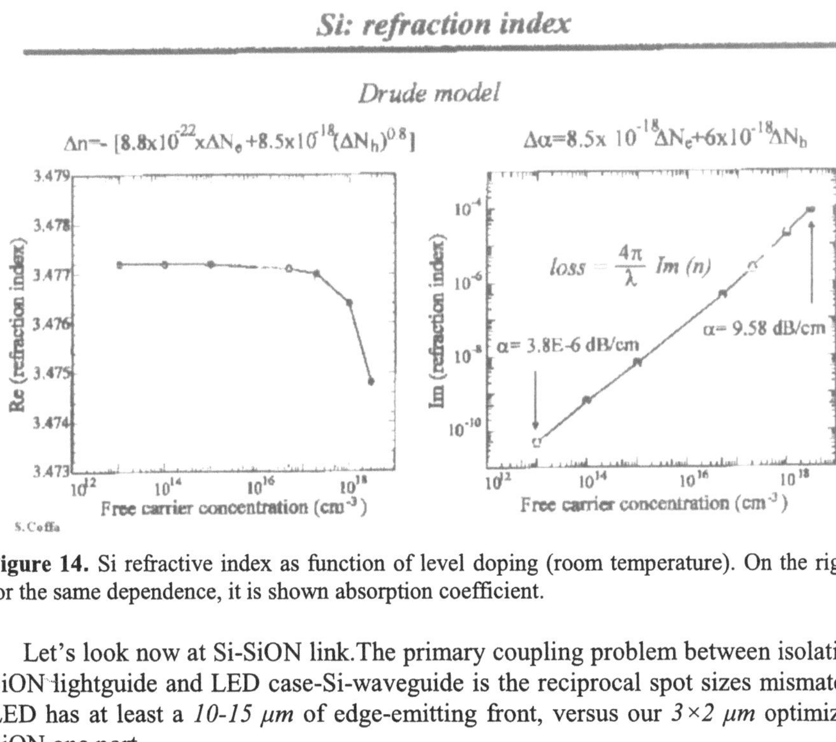

Figure 14. Si refractive index as function of level doping (room temperature). On the right, for the same dependence, it is shown absorption coefficient.

Let's look now at Si-SiON link.The primary coupling problem between isolation SiON lightguide and LED case-Si-waveguide is the reciprocal spot sizes mismatch: LED has at least a *10-15 μm* of edge-emitting front, versus our *3×2 μm* optimized SiON one port.

To solve this problem of ports-mismatching, we can apply a linear *tapering* solution [3] making use of a *multibranched* isolation waveguide: in Figure 6 it is elucidated a possible answer with two isolation lightguides branches delimited by one internal trench.

During work progress, we developed a better solution by two "*semitapering*" (of semifunnels) to three collecting branches and two internal trenches (see below).

6 We use such buried layer to have a planar mirror and then extract a sort of LED case-Si-waveguide. See Figure 11 as regards of the n++ Si lower refractive index than n Si one.

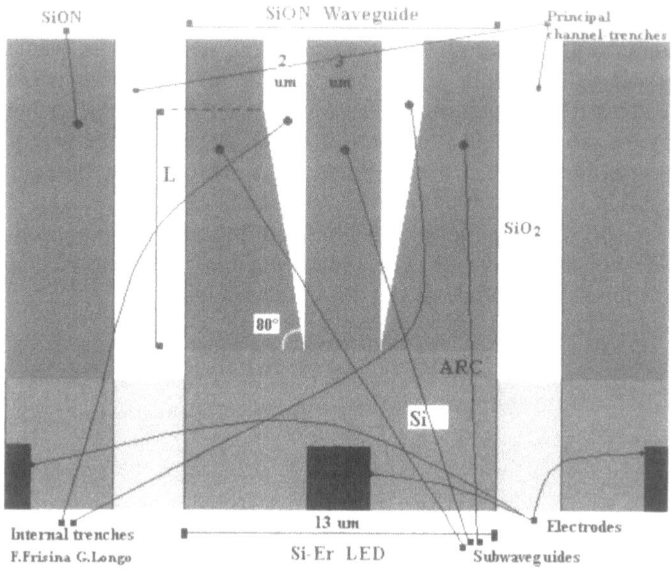

Figure 15. Schematic regarding manyfolded tapering to triple SiON waveguide. It is coupled to a 13 μm front LED (top view joint). It is shown also the ARC (Anti Reflecting Coating).

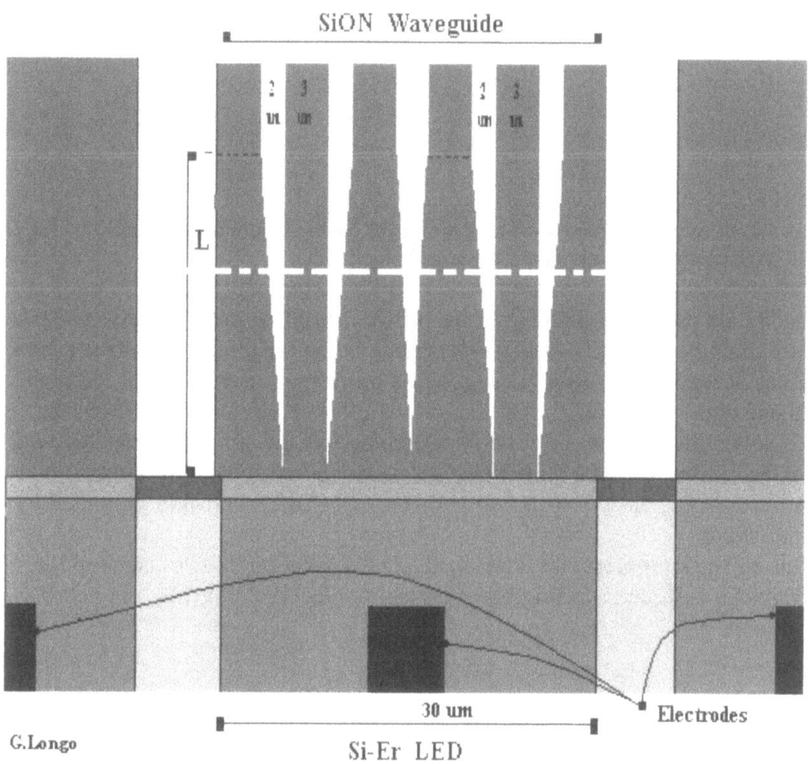

Figure 16. Schematic (not in scale) of six-fold isolation lightguides

With the "semifunnels" configuration of Figure15 we would obtain a higher optical power coupling efficiency because we save two bends and its related losses. In addition we will obtain a superior compactness and a minor masking complexity respect to classical full funnel taperings. Furthermore the central sublightguide will not have any bend.

Note that the "semi"tapering angle (**80°**) is large enough so that incidence angle from normal incoming radiation respect to LED interface is far from the related limit angle $\gamma_L = \textit{arcsin(1.45/1.5)} \sim 75°$. The length L, with a associated tapering port **5 μm long**, will be about **12 μm** (and hypotenuse about **13 um**).

Which are the criteria applied to dimensioning such structure? First, with multi-branching we restricted the ratio $\mathbf{W_{in}/W_{out}}$; now is **5/3 < 2** against the initial **13/3 > 4** , in this way we could putted ourselves in a better condition to choose the tapering length L.

Then we choose L as 10 times our 1.54 μm Er^{3+} wavelength (Ching-Ting Lee et al. 1997); this L = 13 μm is however the worst case, with a loss of about 1 dB/joint for right and left sub-waveguide of Figure 15, while the central straight sub-waveguide do not will suffer of tapering losses.

Nevertheless, we can always fix a longer SiON so that losses rapidly decrease with increasing taper length L (up to negligible losses for L = 1 mm).

It is clear that can be adopted a multiple number of the Figure 15 structure, for obtaining 30 μm, 45 μm, 50 μm ... LED fronts.

In Figure 16 are sketched the six branches-30 μm LED front solution. Here, tapering angles exceed **80°** and **L ~ 25 – 30 μm.**

With a 30 μm long LED front, we achieve a larger photons emitting area and a larger receptive detector area.

It is to point out however that, naturally, if we do not want to lose efficiency in higher frequencies response, when we increase the front, then we must shorten the LED length in a way that junction area is always 600-900 um^2. This avoids making the junction too capacitive.

Note that from tapering structures, there will be *essentially* an optical power transmission, while for modes transfer (e.g., for the fundamental mode[7]), we will confide in a selection of modes by dedicated functioning of our optimized middle SiON waveguides.

To increase further Si-SiON coupling, we can resort to an ARC: the Anti-Reflecting Coating, in the vertical interface between the LED and the lightguide. *In an early prototype stage, to avoid fabrication steps complexities, we will not fabricate it.* For this future step, we could adopt a vertical layer of the same SiON. The material property and the dimensioning of ARC layer must be so that its refractive index is

$$n_{AR} = (n_{\mathrm{Si}} n_{\mathrm{SiON}}) = (3.47 \times 1.5) \approx 2.28$$

and its thickness **s** (see Figure 13) satisfies the resonance condition

---

[7] Our 3×2 microns rib guiding structure, preferably will select the Gaussian mode, but we seen that are permitted also five-six modes of higher orders illustrate in Figure 8.

$$n_{ar}s = (2m+1)\lambda / 4$$

Then, to imitate a rudimentary ARC layer system, it is a very beneficial trick to render, in fabrication phase, the adjacent LED boundary tapering SiON, with a refractive index of **2.28** (*i.e.*, higher than middle waveguide SiON (1.5)), for an extension of **170 nm** = **s** (from last formula for best order (**m** = **0**) or, for the less efficient order **m** = **1**,of **0.51 µm** = **s**.

How can we match the right n_{ar} value? For an idea, we can refer to the physical process data reported in Figure 17 that are referred to Si_3N_4.

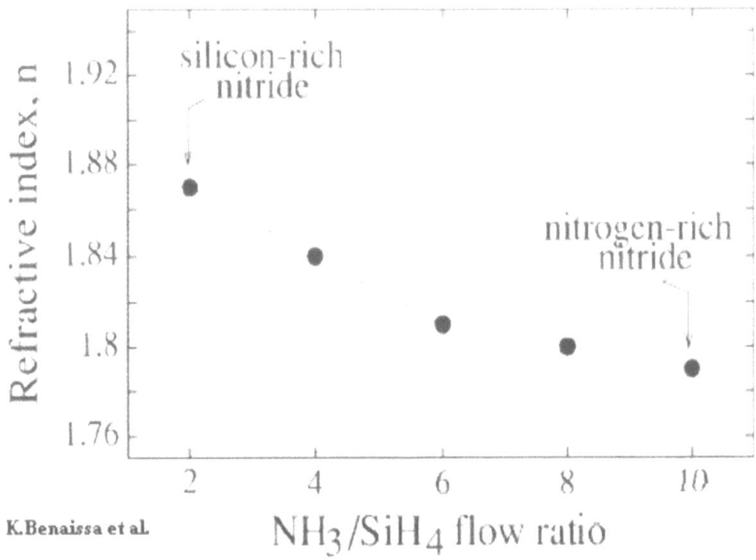

Figure 17. Silicon nitride refractive index variation versus constitutive materials flow ratio deposition (SiON similar)

Concerning our SiON ARC, we first say that it is more advantageous than the in Figure considered Nitride because it reaches **2.008** as maximum refractive index value (Si_3N_4 reach 1.87). It is not the requested 2.28 but is better than 1.5 of middle SiON and could be of decisive rule for a good efficiency of transmitted radiation.

To obtain this high index assessment, we must put down Silicon oxinitride at low-pressure chemical vapor deposition (**LPCVD**) in a hot wall reactor from SiH_2Cl_2, O_2 and NH_3, applying temperatures between 800-900°C and a pressure ranging from 50 to 200 mTorr. The deposition rate is rather slow and can reach 10 nm/min.

To conclude the isolation SiON lightguide section, we present now a portion of it that is far from taperings, standing in proximity of detector connection; the related layout is required to submit it to a optical CAD simulation.

At CNR-IMETEM we have disposition of BBV® optical CAD tools and we are ready to input this data.

Unfortunately, for the work we had two months of real time, therefore, starting far from this point, the results are not ready to be presented. However, we have an idea about our middle SiON performance (to be corroborated by the simulations), and this is shown in Figure 19.

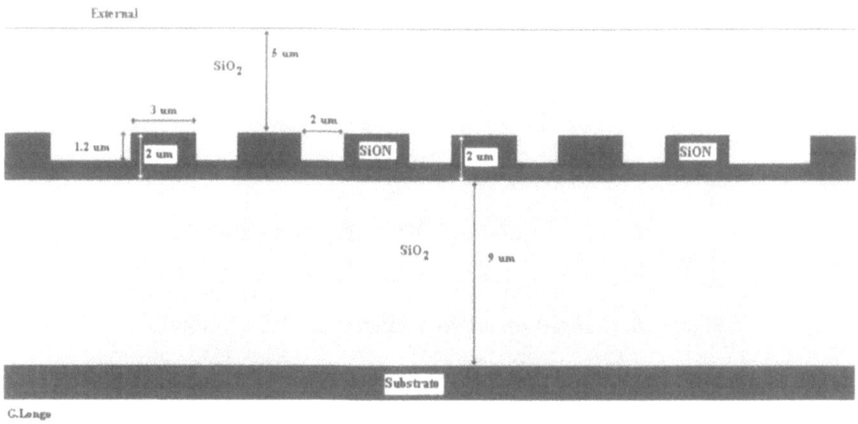

Figure 18. Split isolation waveguide middle-final cross section proposed for optical CAD simulation

Figure 19. Expected (for fundamental mode) six fields collective behaviour (by hand drawn and not in scale)

RIE DBR Improvement

This is a further process step that can be useful to improve both LED efficiency and coupling between SiON and LED. The answer makes use of a Distributed Bragg Reflector (DBR).

As a matter of fact, by DBR we could extract a resonant cavity and so achieve:

i. a higher EL gain at supply equality (see Figure 20);
ii. a narrower frequency selection;
iii. a smaller solid angle emission;
iv. a benefit to the spontaneous light emission by internal-self-reflected light on active medium (RCE, Resonant Cavity Enhancement).

From the circuit point of view, RCE allow to fabricate a LED smaller junction area and then to obtain a lower parasitic capacitance for a higher function speed.

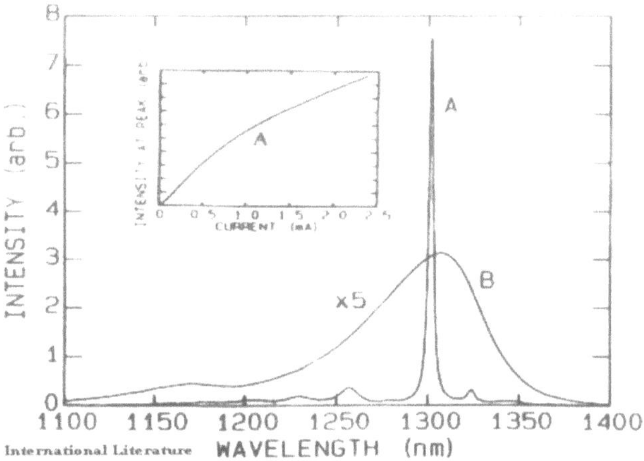

Figure 20. Comparison between ordinary and RCE LEDs EL

Unfortunately, our actual LED is too large to settle it in a cavity. Nevertheless, we will enjoy of reflected radiation by back-DBR action. Our system has planar development; therefore, to arrange some mirrors, we cannot adopt the classical and easier horizontal multilayer stacks. In this circumstance are requested vertical semiconductor micropanels made by RIE (Reactive Ion Etching) as well as in Figure 21. Concerning RIE methodology DBR, confidently we can settle 2-3 vertical Silicon meshes. In the beneath Figure there are 4 mirrors; it is maybe a weak corrective, but we must risk this configuration for reasons which will be soon explained.

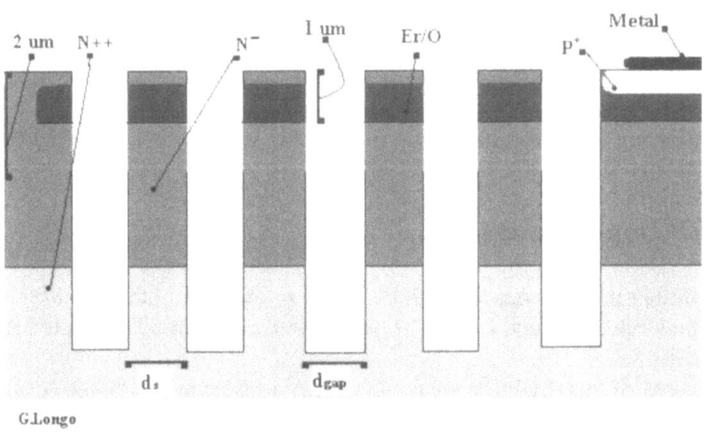

Figure 21. DBR Lateral Cross-Section coupled to our LED. The underneath calculations furnish $d_{gap} = 0.4$ μm (k = 0) and $d_s = 0.56$ μm (m = 3)

As an alternative of the 25-30 layers of multilayer planar DBR, we will use few panels because here we can enjoy high index difference between Silicon and the surrounding medium.

The fundamental dimensions are d_{gap} and d_s, *i.e.*, medium and semiconductor thickness. About them we can say:

i. into semiconductor panel, as it seen in Figure 21, it is convenient to let in Erbium too: perhaps it will can get advantage from a weak electromagnetic pumping;
ii. the best surrounding medium is air, which has the lowest refractive index ($n_{gap} = 1$).

To optimize the mirrors reflectivity we must obey to the general Bragg resonance condition for normal incidence of 1.54 μm (plane) radiation:

$$m\pi = \beta_{gap} d_{gap} + \beta_s d_s$$

$\beta \equiv kN \equiv kn \sin\theta = 2\pi n / \lambda_0$ for normal incidence.

In order to have constructive interference in the air contained between two micropanels, let's fix d_{gap} equal to

$$d_{gap} = (2k+1)\lambda_0 / 4n_{gap}$$

For k = 0 arises

$$d_{gap} = 1.54/(4\times1.0) = 0.39\ \mu m$$

To obtain silicon thickness we will use the previous Bragg condition

$$d_s = (m\pi - \beta_{gap} d_{gap})/\beta_s$$

Then we must calculate the two propagation constants β_m and β_s. They result respectively

$$\beta_{gap} = 2\times3.14\times1.00/1.54\ (um^{-1}) = 4.10\ \mu m^{-1}$$
$$\beta_s = 2\times3.14\times3.47/1.54\ (um^{-1}) = 14.15\ \mu m^{-1}$$

For m = 3 (m > 3 are too inefficient orders), mirrors thickness will results

$$d_s = (3\pi - 4.10\times0.38)/14.15 = 0.56\ \mu m$$

Insisting on our resolution technology power to challenge the situation, we could wool-gather a second order to obtain (m = 2)

$$d_s = (2\pi - 4.10\times0.38)/14.15 = 0{,}33\ \mu m$$

Neglecting the second-order calculation and focusing on more feasible third order, by examining Figure 22, we expect to obtain, at 1.54 μm wavelength, a not exciting 90% of theoretical reflectivity efficiency and 60% of experimental one in a very narrow interval of λ.

Pondering again calculations and simulations results of the next 5 next graphics, we could imagine that, to compensate the value m = 3, we could increase mirrors number from the already enough 3 to 4 and so perhaps improve the global reflectivity at central 1.54 µm wavelength (view the extrapolated Figure 23). Maybe the fourth mirror is a weak solution, but we think that in prototype phase this is a gambler chance to play.

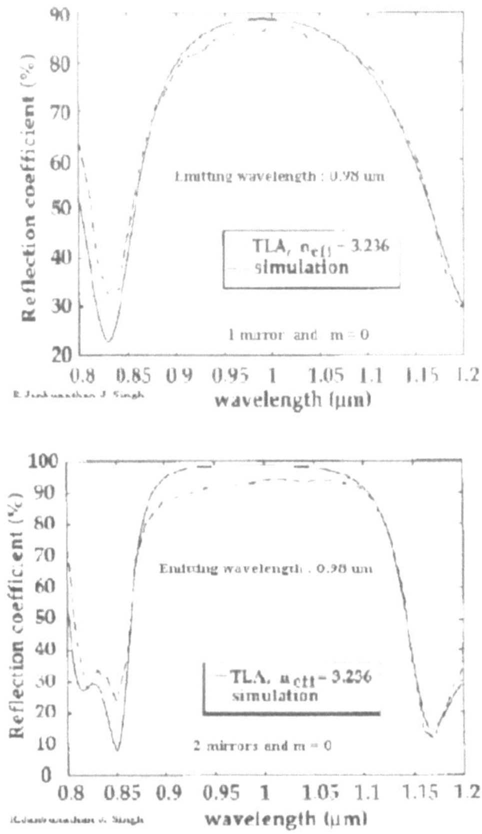

Figure 22a. Theoretical (Transmission Line Analysis) and simulated reflectivity of a RIE DBR. Are indicated order and mirrors numbers.

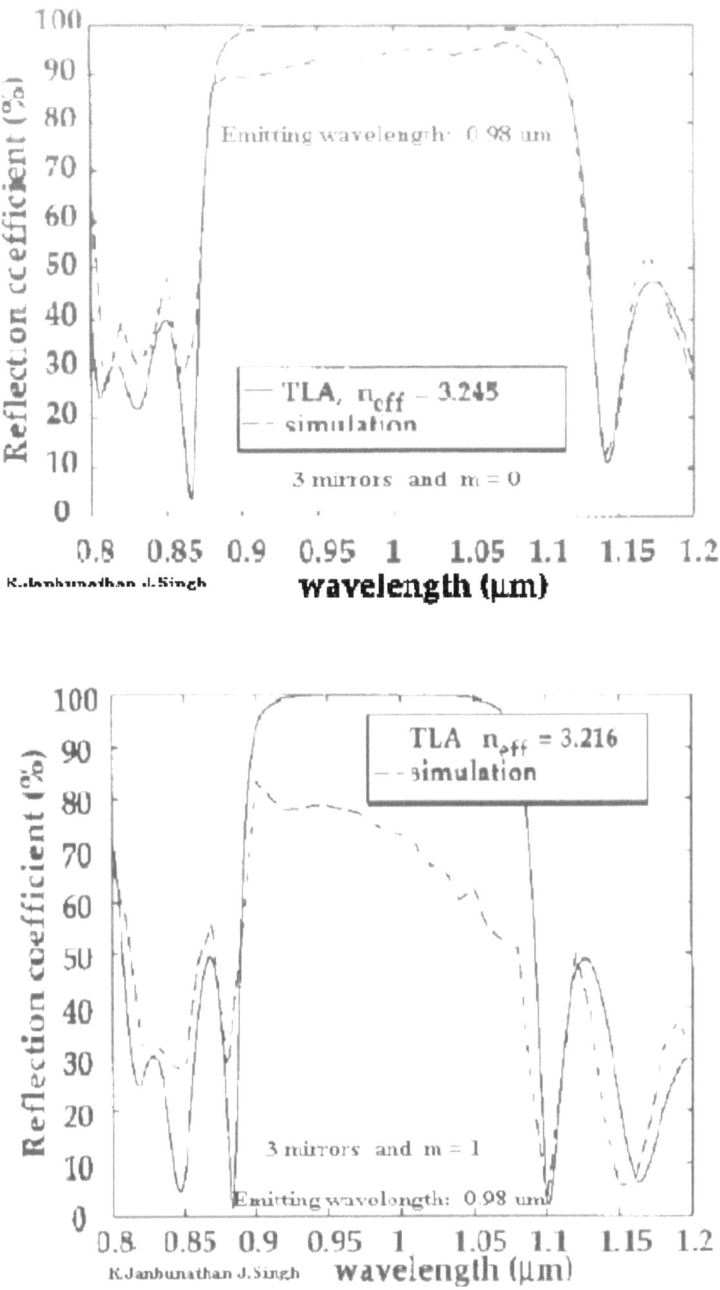

Figure 22b. Theoretical (Transmission Line Analysis) and simulated reflectivity of a RIE DBR. Are indicated order and mirrors numbers.

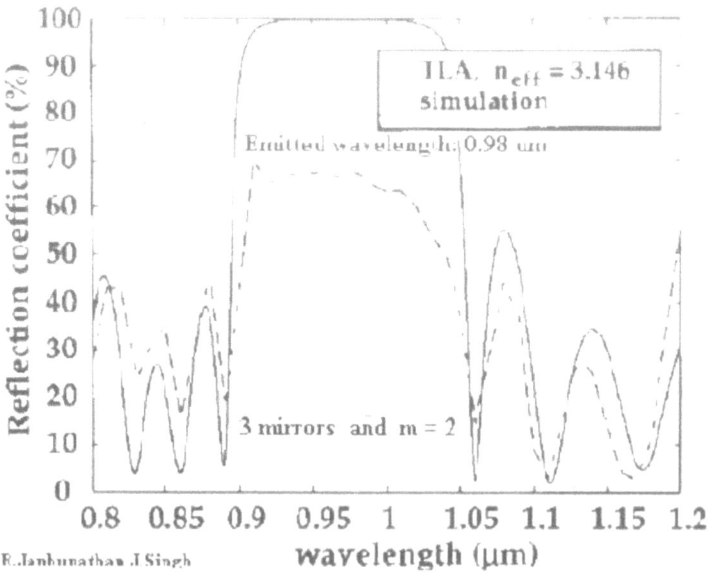

Figure 22c. Theoretical (Transmission Line Analysis) and simulated reflectivity of a RIE DBR. Are indicated order and mirrors numbers.

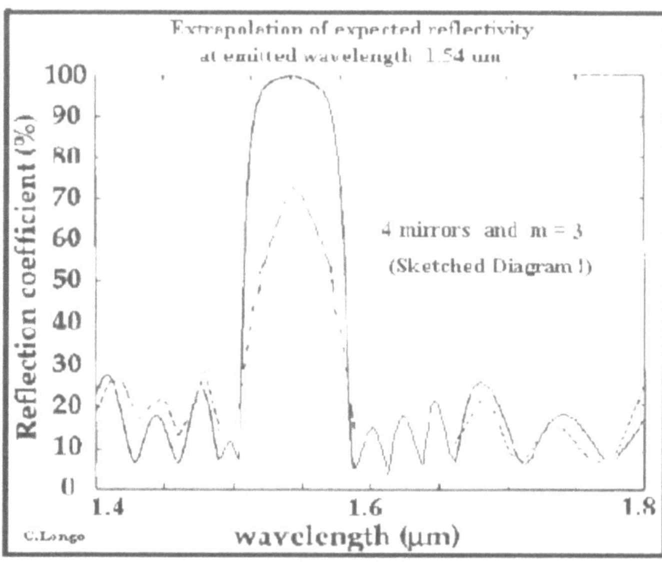

Figure 23. An our prediction outline regarding the here planned DBR behaviour

Detector: Coupling to the SiON Waveguide and Scheme of a Si-Ge MSM Alternative to Schottky

We will first consider the coupling between the SiON and the readily available Schottky infra-red detector.

Relying on our large area of multibranches SiON lightguide, we attached it directly to the detector without any ARC and to improve performance we extended the Schottky junction also to the wall of Principal-Channel-Trenches, with a vertical metal (sidewall metal).

The final right portion (looking at detector front) of photoSchottky, just near the boundary wall of the related Principal-Channel-Trench is shown in Figure 24.

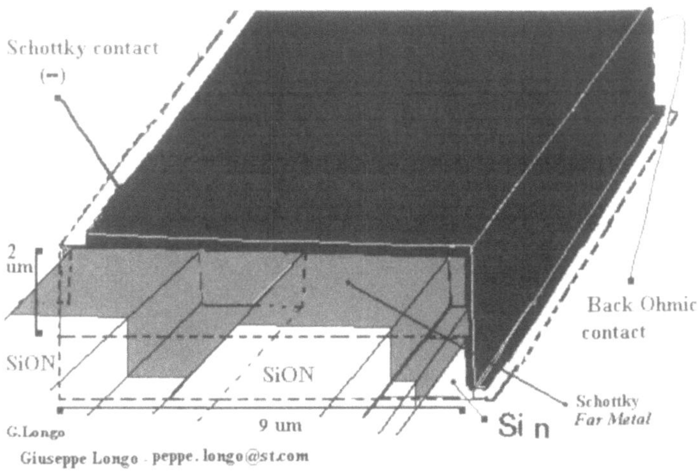

Figure 24. Suggested Schottky (quarter part portion) boundary geometry and simple end-fire coupling with SiON (note the Schottky Far-Metal)

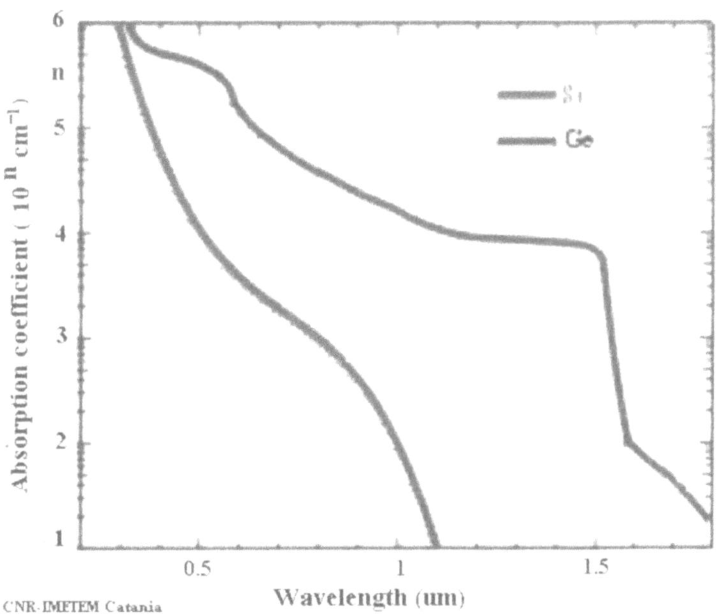

Figure 25. Silicon and germanium absorption coefficient at normal conditions

For our erbium wavelength (0.8 eV) and for silicon (E_g = 1.12 eV), it is demanded that photons must be absorbed by metal; then it easy to *think to extend the Schottky metal layer over a sufficient SiON top surface.* In this way, photons are coupled with Schottky metal by cap face of the final part waveguide. Note the unconventional Schottky metal position, so far from silicon; surely the relative novel layout will need some experimental routes.

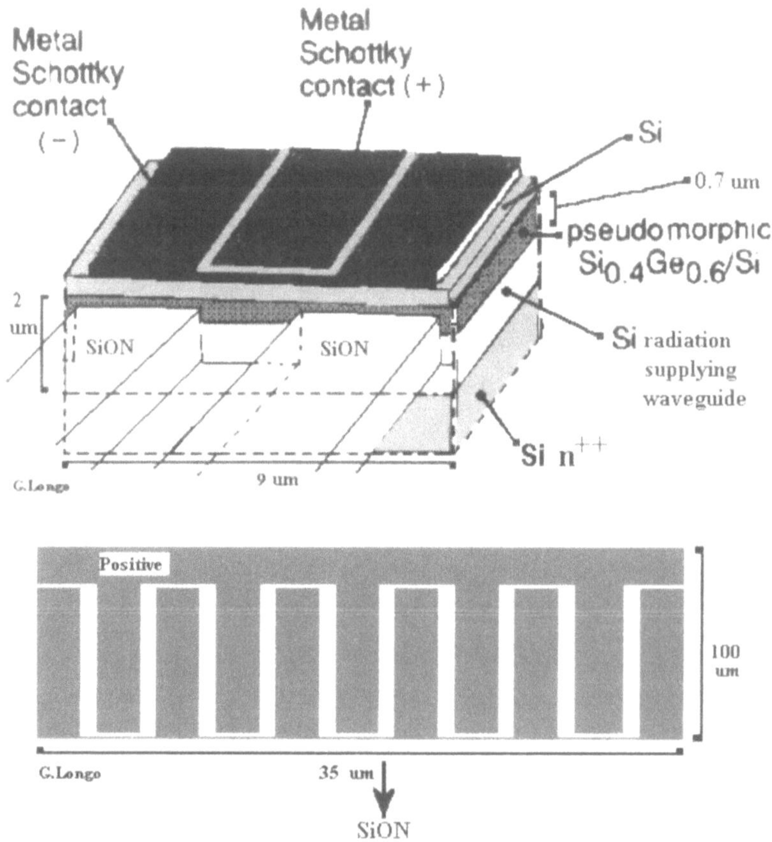

Figure 26. Proposed Si/Ge Detector with its joint from a sixfold multibranches isolation lightguide. Below: electrodes layout.

As an alternative detecting scheme to the proposed "*far-metal*" Schottky detector, it was chosen a Si/Ge superlattice solution. Figure 25 shows that the absorption level of germanium is more efficient than the silicon one for wavelengths higher than 0.4 μm.

At our 1.54 μm wavelength, absorption coefficient of germanium is far more than 10000 times greater of silicon one.

When gemanium is used in heterostructures with silicon molecular beam epitaxy, the related photodetectors extend their response up to 1.6 μm wavelength with a good quantum efficiency (ratio between number of electrons-holes collected at electrodes and number of incidence photons on active area) equal even to 40%.

Now proposed detector is an MSM (Metal-Semiconductor-Metal) with a pseudomorfic Ge_xSi_{1-x} multilayer[8] absorbing region (see Figure26).

MSM is a back-to-back Schottky detector (without any ohmic contact); it works with large applied voltage (at least 10V of bias) in order to completely deplete the MS junction and set the photoproduced carriers at saturation velocity.

Regrettably, a 1-2 microns stair is predicted, although we will be able to use a simple **slab lightguide** thick 1.6 micron instead of the multibranched ridge SiON. The absorbing region is an MBE epitaxial multilayer of 20-30 periods made by 6 nm of $Ge_{0.6}Si_{0.4}$ (on *n* Si substrate) covered by 30 nm of intrinsic Silicon: total height of 0.7 μm. SiGe superlattice will be deposited both on the slab SiON either on the receiving silicon, for a longitudinal length needed for the best detection.

The basis silicon has a waveguiding purpose, to distribute the radiation, longitudinally along the pseudomorphic layer over the receiving silicon (Si radiation supplying waveguide): using Silicon area, we will save free isolation lightguide surface, as is shown in Figure27.

Typically, to obtain a responsivity R of *0.2 A/W* [9] at 1.54 μm, it needed several hundred microns of longitudinal length; dark current is about 600 pA.

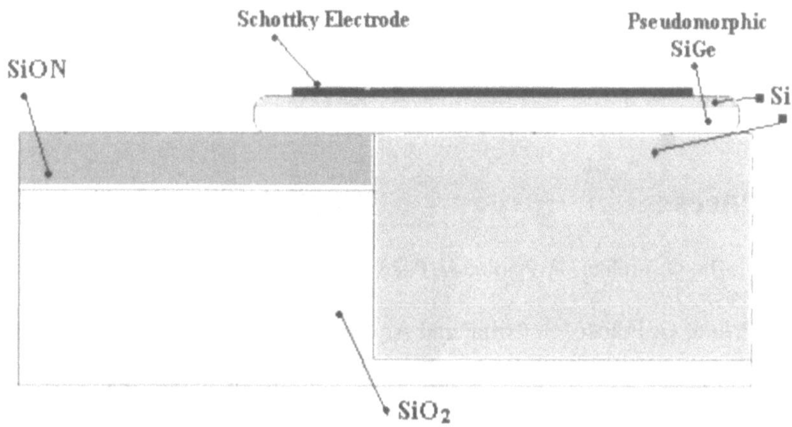

Figure27. SiGe/slab SiON cross section layout instance

As a powerful photodetector, it is convenient to adopt the planned MSM Si-Ge based diode, also because it uses a technology with FET one compatible. Its bulk structure was developed at Technical University of Berlin and before, by H.Temkin et al. [4]. On the contrary, the presented configuration, layout geometry and electrodes planning, are in-progress attempts.

The dedicated interdigited electrodes (which are not in ohmic contact), has not exposure task (also in this case we have a lateral detection) but, meant for our projecting, operates for a symmetrical and short path in favor of photoproduced elec-

[8] The superlattice structure is not mismatched by the 4% superior dimensions of Germanium atoms respect to Silicon ones.

[9] $\mathbf{R = I_{ph} / P_{hv}} = A\eta\lambda$ i.e ratio between photogenerated current and intersecting optical power. *A* is a constant and η is quantum efficiency.

trons-holes pairs. Its shape permits us to obtain a large light sensitive detection area as long as our multibranches waveguide total port.

Finally, note (lower panel of Figure26) the relatively little longitudinal detector extension; in fact normally it can be also 1 mm long: in our case, we should enjoy of our long front waveguide (more than 30 μm) to reach the needed about 3000 μm^2 of sensitive area.

This detection system surely has more complex processing step than Schottky one, but it assure a higher efficiency since offer a larger capable active region.

Conclusions

In this work, a design of a Si-based integrated Optocoupler has been presented. Rationally, as first prototype, we will put into implementation a simple structure with a SiON rectangular parallelepiped as waveguide (the slab instead of the multi-ridge) and the *far metal* Schottky detector, instead of the MSM detector.

Furthermore, as escape route keys or for a successive more efficient transmission structure, additional sophisticated resolutions as multi-branched ridge waveguide, ARC, DBR, Si-Ge MSM detector have been proposed and illustrated for a system combining the devices (LED, lightguide and detector) with novel procedures for light coupling.

References

[1] S. Coffa, G. Franzò, F. Priolo, A. Pacelli, and A. Lacaita, Appl. Phys. Lett. 73, 93 1998.

[2] F. Priolo, G. Franzò, S. Coffa, and A. Carnera, Mat. Chem. and Phys. 54, 273 1998.

[3] Ching-Ting Lee et al., Journal of Lightwave Technology. 15, 403, 1997.

[4] H. Temkin et al.Appl. Phys. Lett. 48, 963, (1986)

[5] M.J.A de Dood et al. MRS Symp. Proc. 422, 219,1996.

[6] A. Polman et al. J. Appl. Phys. 73, 1669, 1993.

MIX
Papier aus verantwortungsvollen Quellen
Paper from responsible sources
FSC® C105338

If you have any concerns about our products,
you can contact us on
ProductSafety@springernature.com

In case Publisher is established outside the EU,
the EU authorized representative is:
Springer Nature Customer Service Center GmbH
Europaplatz 3, 69115 Heidelberg, Germany

Printed by Libri Plureos GmbH
in Hamburg, Germany